STILL FISHIN'

STILL FISHIN'

The BC Fishing Industry Revisited

Alan Haig-Brown

HARBOUR PUBLISHING

Harbour Publishing Co. Ltd.
P.O. Box 219, Madeira Park, BC, V0N 2H0
www.harbourpublishing.com

Cover photo by Alan Haig-Brown. The seiner *Queens Reach* (built by
Allied Shipbuilders, 1980) off Steveston during the 2009 herring fishery.
Author photo by Rick Blacklaws.
Edited by Silas White.
Cover design by Teresa Karbashewski.
Printed and bound in Canada.

Harbour Publishing acknowledges financial support from the
Government of Canada through the Book Publishing Industry
Development Program and the Canada Council for the Arts, and from
the Province of British Columbia through the BC Arts Council and the
Book Publishing Tax Credit.

Library and Archives Canada Cataloguing in Publication

Haig-Brown, Alan, 1941–
Still fishin': the BC fishing industry revisited / Alan Haig-Brown.

ISBN 978-1-55017-467-0

1. Fisheries—British Columbia. 2. Fishers—British Columbia.
I. Title.

SH224.B7H35 2010   338.3'72709711   C2010-900048-X

*To my wife Ananya Surangpimol
and the fishing families of BC*

Contents

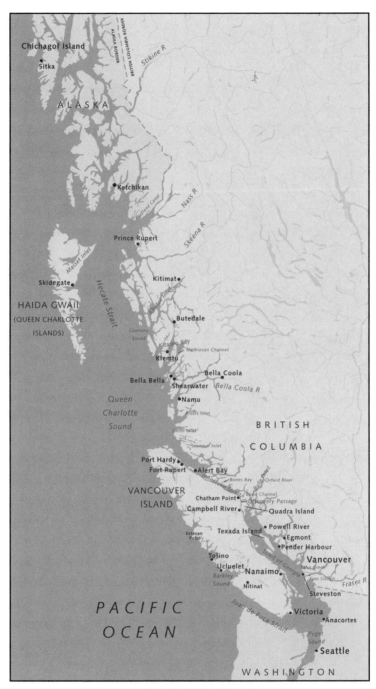

COASTAL BRITISH COLUMBIA

Introduction

As privatization of the resource shrinks the commercial fishery, so also the volume of the fishermen's voices is reduced. This is a pity as these are important contributions in the conversation on the future of our coastal resources. In order to amplify these voices I have interviewed a select group of fishermen and a couple of non-fishing vessel owners. They are a diverse group, from the optimism of twenty-year-old troller Russell Sanderson, to the wisdom of eighty-year-old Hutch Hunt. I have also included two non-fishermen, Randy Reifel and Dan Nomura, in recognition of their roles in controlling licences. As much as the popular press tells of the death of the commercial fishery, the people in this book speak to a future in which commercial fishing can continue to provide employment for those willing to do the work, and provide quality seafood for the consumer.

Kamchatka

End of an Era

In 1926, Menchions Shipyard in Vancouver built five seine boats for Wallace Fisheries' pilchard-and-salmon-processing plant in Barkley Sound on the west coast of Vancouver Island. These fine, big boats were christened with the unimaginative names *W#7*, *W#8*, *W#9*, *W#10* and *W#11*. That same year, just along the waterfront from Menchions, Harbour Shipyards built the seiner *Kamchatka* for Ryotaro Kita. A year later, up north on Cormorant Island, a shipwright named Lyunggren built a seiner for Julian Olney, who named her *Alert Bay* after that island's famous fishing town. These boats were part of a fleet of well over 150 seine boats built in British Columbia in the five years between 1925 and 1929. Japanese-Canadians built boats to fish herring, Euro-Canadians built to fish pilchards, and everyone built to fish or pack salmon. Good fishing, along with a booming forest industry and rum-running into Prohibition-era USA, combined to pump a lot of money into coastal communities.

Immigrants from Norway, Japan and Croatia arrived with strong fishing traditions and a hunger for new fishing grounds. US and British investors wanted a piece of the untapped resources of BC. A 1927 issue of *Harbour and Shipping* magazine described the versatility of seine boats on a coast with few roads:

> The development of special fishing craft in connection with seining operations on the British Columbia coast has been an outstanding feature of the last two years. . . . As seine boats built for the salmon industry are also suitable for seining herring and pilchard, and can be adapted for halibut fishing, besides being useful for towing and freighting, there has been a good demand for this class of boat, and most of the local shipyards are busy on these craft again this spring. . . . These fishing craft are heavily constructed of fir, with steam bent oak frames. Formerly they were all equipped with heavy-duty gasoline engines, but in recent years oil engines are finding favour.

The Depression ended the boom and many boats sat idle through the 1930s. In the 1940s, many of the boats and their men were pressed into military service, patrolling the coast armed with rifles against enemy submarines. In the 1950s, the boats were repowered and hydraulics were added first for power blocks (and later in the 1960s for drums, and in the '70s and '80s for tilt sterns and vanging winches). The little wheelhouses became crowded with electronics: first sounders and phones, then radar, loran and plotters. The pilchard disappeared in the late 1940s, the herring were fished out in the 1960s, but the salmon kept the boats working through the summer months. The

The *W#9* in the early 1980s was approaching 60 years of age but still in fine fishing form.

1950s and '60s were good for seining and for the coastal communities. Many of the boats from the 1920s were now fished by First Nations people from communities with long histories of sustainable resource harvesting. These were boats that could support several families on four sets of the net per day and still provide employment for ship-wrights and other maintenance in the off-season.

The 1970s saw the first of the limited-entry plans, re-sulting in a fixed number of 530 seine boats on the BC coast. This was an increase from under 400 boats as the plan allowed several small gillnet licences to be combined on a single new seine boat. Once that avenue was closed, the only way to build a new seiner was to retire an old boat. The prices of boats were under $50,000 before the limited entry shot up four times and higher with their newly attached licences. The companies with large fleets profited from the spectacular appreciation in boats and li-cences. At the same time, government worked to enhance sockeye runs and the boats made money.

Through the 1980s, many 55-to-65-foot wooden seine boats with 14-to-16-foot beams were retired and their licences were put onto new deep draft, high free-board aluminum, steel and fibreglass boats of the same length with 22- and 23-foot beams. Tilt sterns were added to bring the fish aboard quickly. With a salmon licence, these boats were costing upwards of a million dollars and routinely making eighteen or even twenty sets per day to achieve the volume of fish necessary to service their debts. Boats that could pack less than 50 tons of herring were replaced with boats of the same length that could pack twice as much. The less well-built or well-cared-for old boats were sold to less-developed fisheries like sea urchin or geoduck diving. Some were converted to pleasure boats and others mouldered at company docks. As the fishing

The *W#10*, one of a series built at Menchions in 1926 for the pilchard fishery, awaits demolition. The fine double-decker *Renown*, built at A.C. Benson Shipyard on Coal Harbour in 1957, is still on the Canadian register.

companies entered the salmon farm industry, a few of their boats found work tending these marine feed lots. British Columbia Packers records show that in 1984, Port Hardy fisherman George Hunt skippered the *Kamchatka* while the *Twin Islands* (ex-*Alert Bay*) was fished by Dan Smith and the *W#10* by Larry Holmes, both out of Alert Bay. The BC Packers roster for that year included just over 200 boats owned or contracted to the company.

The 1990s brought another federal plan to reduce the size of the BC salmon fleet. Once again picking up the cliché of "too many boats chasing too few fish," and heralded with political rhetoric of the "steak in every fisherman's frying pan" variety, many fishermen said it was more like "a stake in the heart" of coastal communities. It tended to be these communities, with their lower costs of living, that supplied the skippers and crews for many of the older

The *Kamchatka* moored with the fleet at the BC Packers old Celtic shipyard on the Fraser River's North Arm. The seiner was built at Harbour Shipyards in 1926 for Ryotaro Kita. The *Bates Pass* was built at Park Shipyard for BCP.

boats. Under the Mifflin plan, the licences for some of the 530 seine boats were bought back by the government and retired while others had their licences "stacked" so they could fish the two areas into which the coast was now divided. The Coastal Communities Network, supported by the United Fishermen and Allied Workers Union as well as a number of environmental and municipal organizations, argued that it was the most sustainable fishermen fishing the relatively debt-free, less capital-intensive old boats who found themselves out of work when licences were stacked. Each boat retired left four or five crew members on the beach.

The Mifflin plan drove up the cost of licences again, as each boat needed a second licence at an additional $200,000 or more if they wanted to continue fishing the whole coast. Those who owned boats found themselves unable to afford a second licence. Worse yet, those who fished older company boats found themselves on the beach while the company stacked the licence from their vessel onto a more modern seiner. Fishing companies that had carried fishermen on credit for years had less motivation to service these debts. The companies also continued to close the shipyards and net lofts that once supported the larger fleet. Waterfront land was now worth more developed for residential use by what one BC union leader, Jack Munro, has called "goddamn cappuccino-sucking, concrete-condo-dwelling yuppies."

The fleet reduction had little if anything to do with fish conservation. It was, according to salmon fisherman and economics professor Don Pepper, a further illustration of the inability of the Canadian Department of Fisheries and Oceans to define a goal in the management of the west coast fishery. Pepper kept in touch with the industry by going out to the grounds with his old friend and skipper, Byron Wright. Until health forced him to sell his boat and licences shortly before his death in 2008, Wright was one of a diminishing number of truly independent owner-operators. He got his start as a seine deckhand out of Alert Bay on a boat very much like the *Kamchatka*. As the owner of a modern aluminum salmon and herring seiner in the late 1990s, Wright told me of management problems caused by area licensing in the British Columbia herring fishery:

> The result has been a systematic transfer of the fishing privilege from individual vessel-owners to large fishing corporations which both numerically and functionally control the operation of the industry from fishing to marketing. This process of consolidation has made management of the fishery political, as the major companies continue to press for further control. This is understandable, as they wish to protect their interests, but this may not be in the best long-term interest of the industry and certainly not in the interest of the few remaining independent operators.

First Nations fisherman George Hunt, who skippered the company-owned *Kamchatka* for seven years in the 1980s, was given work under the Mifflin plan's retraining provisions: "We were supposed to learn some other trade, but first they had us picking up garbage and making trails. The only good thing was that we took a navigation course, but now I can't afford a boat."

Just as the old fishermen and their children were left on the beach, the major fishing companies found there isn't a lot of demand for 60-foot wooden boats in the pleasure boat world, so many of these still-sound vessels have been pulled up for demolition. In September 1997, BC Packers stripped the machinery, winches and brass work from the *Twin Islands* (ex-*Alert Bay*), the *Kamchatka* and the *W#10*— after a combined 212 years of working the coast. Early in 1998, I stood at the former Nelson Bros. Queensborough Shipyard and witnessed the destruction of the *Kamchatka*. As a backhoe shovel's hydraulic jaws crashed their way through the fo'c's'le planks, the fine old-growth edge-grain fir cracked to reveal black bilge stain that penetrated only a short distance into the rich red of the wood. Oak

The *Kamchatka* under the hoe at the BC Packers, former Nelson Brothers, Queensborough Shipyard as finely shaped old-growth wood yields to the machine.

frames, showing only a little age above the waterline and still sound in the curve of the bilge, splintered in spite of their strength. The planks parted from the bow stem and the graceful curve of the forefoot around which long-gone fishermen had pulled the old manila lines with their Spanish corks and cotton web. The shovel's teeth sank into the keel, tearing away the oil-stained keelson. The keel that had no doubt withstood the occasional grounding would not break under the steel jaws that clutched and tore it from the mess of shattered planks. The starboard side guard bent, then separated from the stem and regained its purposeful shape.

The machine won, of course, and in the end the seventy-year-old boat was reduced to splinters. Was this also a victory for those who complained of "too many boats chasing too few fish"? In strict terms of "catch per unit of effort," meaning return on investment, the answer to this question would likely be "Yes." But taking in employment and families supported by the fish our streams and seas produce, the answer is clearly "No." Fewer boats means fewer jobs at sea and in the shipyards, where modern aluminum boats require less maintenance. (My first job in the fishing industry, in 1960, was the pre-season painting of a wooden seiner only fifty yards from where I watched the destruction of the *Kamchatka*.) Perhaps even more significantly, the fewer jobs that do exist on a corporate-owned boat employ fishermen with less of a sense of responsibility for the stewardship of the resource. Corporate owners don't tend to belong to salmon-enhancement or stream-keeper societies. They don't attend off-season meetings to try their damnedest to help bureaucrats understand the ways of the fish. Rather, accountants and lawyers tell them whether they will make

One of the Wallace seiners in Alert Bay in 2009, and looking like her days are numbered.

Built in 1928 by Yasuhiro Nakade at Steveston for Jisaburo Kasho, renamed for the race horse, the fine-hulled *Sea Biscuit* dies on the beach at Alert Bay in 2001. Note old residential school in background.

a greater return on a fish cannery that employs a couple hundred people or if they should close it and develop the land for condos—or, in the case of one major processor with long-closed up-coast canneries, to clear-cut log the properties. Will more salmon make it to the spawning grounds because these old boats no longer exist? No, the new boats will have to catch more fish to pay for their additional licences and greater capital costs. Will the few remaining boats be able to pay higher taxes to support salmon enhancement? No, they will deduct from their income the dead fish whose dollar values go to bankers eating tax-deductible lunches of fine salmon.

Like the splintered *Kamchatka*, the Roaring Twenties and affluent fifties are behind us now, but many of British Columbia's fishermen say it isn't too late to return fishermen to an active role in the fishery. A limited-entry approach, like vessel quotas in other fisheries, has a place in managing the salmon fishery. In Alaska, the state constitution declares the fishery open to everyone, though with a 1972 amendment requiring limited entry for conservation and prevention of economic distress. But also in Alaska, there is a "use-it-or-lose-it" provision that obliges the licence-holder to be aboard the boat when it is fishing. Bruce Twomley, chairman of the Alaska Commercial Fisheries Limited Entry Commission, explained in the late 1990s that "The law defines these as revocable privileges that cannot be held by a corporation." The result seems to be that salmon-fishing boats are invariably owner-operated.

Critics of the Alaskan system point out that it does not preclude an owner-operator from borrowing money in return for a promise to deliver fish to a given processor. Because the permit itself cannot be used as security for a loan, unauthorized loans with permit brokers or processors or investors would be grounds for revoking

Retired from the sea to become a sign, and facing the ignoble destiny of rot.

the privilege to fish. "If someone lies to us, we can re-voke," stated Twomley. "If someone other than the fisher-man is trying to own the licence, someone usually blows the whistle." The key element typically missing from Canadian limited-entry schemes is the owner-operator provision. Billy Griffith, veteran of forty-two seasons skip-pering salmon seine boats, maintains that "Limiting li-cence ownership to bona fide fishermen would keep the cost of licences down and allow low-impact sustainable fishermen a continuing role in our fisheries and our coast-al communities."

The *Kamchatka* and the Wallace seiners may be gone, but the salmon and herring and other fish can return to the coastal seas. Through government policy changes such as licensing fishermen rather than boats, the fishery could be given back to the people of the coast. Young men and women could once again stand on the shore and imagine themselves in the fishing fleet returning to port. I per-fectly recall my own admiration of 1950s seiners coming into the breakwater at Campbell River, brailers swinging proudly from the booms to declare success. When I did go fishing, it was on a smaller wooden seiner with a captain who was proud to provide for his family. We never made much money, but that boat and the people on it taught me lessons I have followed and enjoyed all my life. This deep and rewarding human connection to the waters and the land could be possible once again in a new millen-nium of Pacific coast fisheries management.

2

BCP 45

Heritage Boat and Heritage Skills

You can tell a nation's values by how it shows respect for its heritage. When I travelled the west coast of Norway in 1989, I was enthralled by the number of well-maintained older wooden fishing boats in places like Trondheim and Kristiansand. One night, rather late, I came out of a bar that was upstairs in a sixteenth-century Hanseatic League warehouse on a pier in Bergen lined with similar buildings. Just across the pier, several fishermen were mending holes in a large, strung-out trawl net. Their large offshore trawler was moored there along with others. The fishermen stopped work for a moment to talk with the revellers pouring out of the bar.

This is a huge contrast to British Columbia where Vancouver's False Creek, once an important working waterfront, has been dolled up with pretty stone walkways where residents and tourists can stroll in the summer sun or winter rains. There are no dirty fishing boats or fishermen to be seen. The rock walls where ships and barges

A well-displayed old fishing boat at the maritime museum in Oslo where fish and fishermen are respected.

once moored have been deliberately shoaled so there is no possibility of any vessel other than a rowboat approaching the fine people who walk there. The area under the Georgia Viaduct, where old men lived in equally old trollers, has been filled in to make space for expensive condos. The owners of the condos complained so much that the last of the live-aboard boats, anchored far from the shore, were driven from the Creek in 2007.

The treatment of the old False Creek fishermen is repeated in most of British Columbia's maritime museums. Norway again provides a contrast. The maritime museum I visited in Oslo has spectacular displays of Norwegian vessels, including the polar sloop *Gjoa* moored there. But the building that houses the "Coastal Norway" exhibit captured my attention. Not only were there traditional rowboats from the country's varied coastal regions, there was also a full-sized fishing boat cut away to show the engine and keel. It caused me to reflect on the differences between Vancouver and Oslo. The majority of Oslo's people come from coastal villages. Even the businessmen I met had immediate ancestors in villages and hundreds of years of history. Very few people that one meets in Vancouver have ever even been up the BC coast. To get a licence for a fish farm in Norway in 1989, one had to prove a family history in marine work in the region. I understand that the corporate world has since taken over, but such a strong appreciation of local history wasn't even considered in BC. It has been said that to gain recognition in the Vancouver Maritime Museum, a boat needs to have been built in the United Kingdom—a bit of an overstatement but compared to Oslo it has a kernel of truth.

Fortunately for British Columbians there are some positive developments in places like Steveston and Campbell River. The Britannia Heritage Shipyard has been struggling with limited budgets to preserve some significant vessels and buildings from BC history. The rebuild of the Fraser River gillnetter *Silver Ann* in the Atagi boatyard is remarkable on a number of fronts, not the least of which is that the boat was the last wooden gillnetter built there. We can all hope that it will be followed by similar projects so those walking the Steveston trail system on that corner of Lulu Island can continue to see master craftsmen like Colin Duffield work magic with hammer, plane, caulking tools and paintbrushes. As a piece of coastal heritage, the beautiful *Silver Ann* will impress and bear witness to a time gone by but also to the living fisheries of the future.

Another Steveston-built boat, the *Soyokaze*, has been enshrined in a permanent display alongside the Museum at Campbell River. Built at the Kishi Boatworks in about 1939 for Quathiaski Cove resident Shigekazu Matsunaga, the 37-foot boat was configured as a cod boat. With a fish hold located well forward, the hull had holes to allow water to circulate through the hold to keep the cod alive until delivered. An aft cabin provided additional accommodation separated from the live-well fish hold by a watertight bulkhead. Another watertight bulkhead forward separated the fish hold from the wheelhouse and the engine room with its 8-10 Easthope engine. These little boats were ubiquitous throughout much of the history of the BC cod fishery. While the little wooden double-enders are gone, fishermen such as Fred Hawkshaw of Prince Rupert continue to earn higher prices by landing and selling high-quality live fish from his modern fibreglass boat.

The *Soyokaze* represents an important piece of the racist history of the BC fishery. Just two years after she was built, the government confiscated the *Soyokaze* along with 1,200 other vessels. The Matsunaga family endured

The *Silver Ann* looking tired and resting on a cradle at Annieville in about 2004. Ironically, the sign declaring the Fraser to be "the world's greatest sockeye salmon river" has been allowed to fall down.

Built in 1969 by Sadjiro Asari for George Osaka in this same shed on the Britannia Heritage Shipyard site, the *Silver Ann* was rebuilt and finally relaunched in 2009.

The restored *Silver Ann* returns, like a salmon, to her native Fraser River waters in 2009.

The Matsunaga cod boat *Soyokaze*, beautifully restored and displayed at the Campbell River Museum.

the war and internment but never forgot their little boat. When finally allowed back to the coast they tracked down the cod boat and repurchased her. Mr. Matsunaga continued to fish the boat until declining stocks closed the local cod fishery. In 1999, the Matsunaga family donated the boat to the museum, where it was restored and installed in a proper facility with due ceremony.

Campbell River has continued to recognize its commercial fishing heritage. A program that recognizes fishing people and their boats has been completed at the community's Maritime Heritage Centre. Located adjacent to the commercial fishing dock where I walked the floats and admired the boats as a child in the 1950s, the building is both a museum and a community meeting place. When I visited there in 2006 I realized two of British Columbia's maritime pioneers had spent a good bit of time together over the previous few years. Buford Haines, boat builder and fisherman, had been rebuilding the classic salmon seiner *BCP 45*. Both the man and the boat had decades of work on the BC coast. Born in 1924, Buford built his first rowboat as a thirteen-year-old at his parents' home at Heriot Bay on Quadra Island. The *BCP 45* was born and went directly to work as a salmon fishing boat in 1927.

Buford Haines' father was a steam engineer in the logging industry, but Buford's interest in building rowboats prompted his dad to send him off to Vancouver to work with Bert Benson at his family's Coal Harbour shipyard. It was 1943 and instead of the usual wooden tugs and fishing boats, the Benson yard was building 110-foot British-designed Fairmiles and 118-foot mine sweepers. It was a good training program and Buford was a quick learner. Starting as a "helper," he was promoted to "improver" within four weeks and was given his own helper after three months. Bensons produced four of the eighty Canadian Fairmiles built in the war years. In 1945, Buford went back up coast and used his new skills to rebuild a small fishing boat that he used to troll for salmon. This set a pattern of building boats in the winter and fishing in the summer. In a two-decade period between the 1950s and '70s, he built twenty boats, mostly about 40 feet, in as many years. Some he built on contract for customers, others he fished for a time and then sold.

In 1955, Buford worked at Quathiaski Cove on Quadra Island, putting a new cabin top on a seine boat belonging to British Columbia Packers. It was the *BCP 45*. Little did he realize that a half century later he would again be changing that same cabin top. This time he was working in the newly built Campbell River Maritime Heritage Centre building across Discovery Passage from Quathiaski. The Centre was built when the 53-foot *BCP 45* was in search of a home. The boat was one of five built to the same design by BC Packers in the roaring good times of 1927. The BC

The Campbell River Maritime Heritage Centre.

salmon and pilchard fisheries were doing very well and the fishing company built two of these boats at their own Celtic Shipyard, and another three at Vancouver's Burrard Shipyard. The *BCP 45* was one of the latter, and one of only two that had survived (the other has been converted to a pleasure boat).

To appreciate the importance of seine boats like the *BCP 45* is to appreciate the last century on the BC coast. Throughout much of the twentieth century, Natives and immigrants living in small coastal communities, logging camps and fish canneries, populated the coast. Everything had to be moved by boat. The salmon season even then was relatively short and confined to the summer months. The west coast purse seine boat was highly adaptable to other fisheries, but it could also serve as a camp tender for the logging industry in the off-season. The Vancouver marine periodical *Harbour and Shipping* noted in April 1927 that "Last year there were about fifty-five new seine boats built on Burrard Inlet and the Fraser River." Constructed in the midst of that grand era, the *BCP 45* is a remarkable testament to the quality and practicality of the versatile vessels that have occupied such a significant role in BC maritime history. For much of her life she remained under the ownership of the British Columbia Packers Company and was skippered by many well-known captains. Most of these captains were from the Cape Mudge or Wewaikai Band of the Laich-kwi-tach First Nation, whose traditional fishing territories are in the salmon-rich waters of Johnstone Strait just north of Campbell River.

The best thing one can wish on a boat is that it should have an uneventful career. As the *BCP 45*'s sisters succumbed to fires and sinking, the little seiner just kept at her work. The first time anyone in Canada outside of the west coast fishing community noticed her was in 1958, when her picture appeared on the cover of an eastern newspaper. The photo showed the boat, captained by Mel Assu of the Wewaikai people, off Ripple Point in Johnstone Strait. It was during the legendary 1958 sockeye season in

Pulling the nets by hand in Johnstone Strait. Seine fishing in the late 1940s.

which fishermen earned an ordinary worker's yearly pay and more in a few intense weeks. Even for boats, fame begets fame and in 1973 the image was chosen to grace the Canadian five-dollar bill. In 1986 the boat's owner, Capt. Allen "Ollie" Chickite, from the Wewaikai Band, was invited to bring her down to Vancouver for display at Expo, where the nearly sixty-year-old vessel proved a popular exhibit.

As mismanagement of the salmon fishery led to fleet reductions in the 1990s, Chickite decided to retire the aging seiner that had served several generations of fishermen. With a good understanding of the *BCP 45* and other seine boats' significant contribution to BC maritime history, he donated the vessel to the Vancouver Maritime Museum; unfortunately she did not fit the museum's understanding of its mandate. Alongside the

The *BCP 45* was donated to the Vancouver Maritime Museum by Ollie Chickite after being on display at Expo 86 but in the end the museum's curators didn't want her and she travelled back up coast to the Campbell River Maritime Heritage Centre.

museum's featured display, the *St. Roch*, the first vessel to circumnavigate North America, the honest little hard-working boat just didn't measure up. The years of toil and modifications, including a massive hydraulic drum in place of the original seine table, rendered her an expensive conservation proposition, and showed her to be poor company for the classic tugs, sailboats and air force crash boats favoured by the museum's board for their limited moorage space.

The people of Campbell River, led by their local Rotary Club, showed more interest in taking on the *BCP 45*. They raised the funds and built a Maritime Heritage Centre to house the boat and form a centre for the area's rich marine culture. In 2002, Capt. Chickite once again took command and ran the *BCP 45* the twelve-hour trip up to Campbell River where she had spent the better part of her life. The beautiful building was prepared on the shores of Discovery Passage within sight of the Cape Mudge village on Quadra Island.

When wooden fishing boats are discussed in Campbell River, the name Buford Haines will come up sooner rather than later. I can remember as a small boy speaking with a child's authority to tell someone "That boat over there is being built by Buford Haines." Although I was merely repeating what I had heard, I was also experiencing an important moment in history. When I visited the Maritime Heritage Centre in 2006, Buford recalled that when he was asked to take on the restoration job he suggested it would be easier and more cost-effective to build a replica from scratch. But the community had an attachment to the *BCP 45*, so he was asked to help patch up the aged craft with the hope that a coat of paint would preserve her. "If you want to sweep it out of the building in five or

Buford Haines was finishing up the *BCP 45* project when this picture was taken in 2006. He died shortly after finishing the job.

Fully restored at the Campbell River Maritime Heritage Centre in 2009.

six years, that will do," he told the organizers as he dug into the rotting timbers.

Buford agreed to take on the job if he could do it his way. Having built an impressive list of wooden, fibreglass and steel boats over his lengthy career, he understood that committees don't build good boats. Given a free hand and with the help of some dedicated volunteers, Buford totally rebuilt the little seiner in a manner that would allow future generations to see a fine example of the boats that helped develop the BC coast. Originally built with Douglas fir planking on bent oak frames, Buford primarily replaced the old wood with beautifully resinous yellow cedar donated by local logging companies. He formed the new frames by cutting 2-by-5/16-inch strips of cedar and then laminating them six deep. Before the glue had dried, he drove them down as sisters beside the old rotted oak frames, allowing the laminated frames to bend smoothly

The *BCP 45* was returned to the way she would have been on the Canadian five-dollar bill in 1958, with table seine and power block.

around the bilge and fit snugly to the keel. With the new frames in place, the crew replaced the old fir hull planking with newly finished edge-grain yellow cedar planking. Similarly, the bow stem and a forward section of the keel were fabricated and installed in place of most of the old wood. The stern of a wooden purse seine boat was built from massive timbers shaped and bolted together, providing the good buoyancy aft that the heavy net and seine table required. Here also, Buford replicated the old with bright new yellow cedar. Instead of caulking a vessel that would never return to the water, the crew made thin, wedge-shaped wood strips to fill the spaces between the hull planks.

By May 2006, the hull was largely completed and a new combing was being fabricated for the fish hold. The deckhouse, with its little galley, pilothouse and flying bridge—so important when the skipper had only his eyes to see the fish—was being restored. The vessel's last engine, a 165 hp Detroit diesel was being refurbished for external display, while a duplicate of the boat's second engine, a BC-built Vivian, had been located, fixed up and installed in the engine room. In the 1950s and '60s, seine boats became increasingly mechanized as hydraulics were introduced. Larger engines were needed to pump the hydraulic oil. Additional weight was added with large drums to wind the nylon net replacing the wooden table on which cotton nets had been piled by hand. The *BCP 45*'s original deck winch for pursing the net with its twin capstans had been replaced but was recovered, refurbished and reinstalled.

Buford and his volunteers completed the restoration shortly before he died in the spring of 2007. He was eighty-three and the *BCP 45* only slightly younger. Today and in the future, the public and schoolchildren can come and see the tide-twisted waters of Discovery Passage as well as the fine hull of one of the many seine boats that worked the salmon-filled waters. Few of the visitors can fully appreciate the span of time from the 1920s to the twenty-first century, and all the fish that the boat caught. Similarly few can fathom the deep understanding of wooden boats that Buford Haines had accumulated in the years since he built his first 8-foot skiff. But in the years to come, all visitors will be able to witness a vestige of the era when wooden boats and salmon defined the BC coast. Not only does this serve as a memorial to the wooden-boat builders and fishermen of the coast, but the *BCP 45* stands as a promise of the continuity of the coast's commercial fisheries.

3

New Venture

Returning to the Coast

When Jamie VanPoele sits across from the fireplace and turns on his big flat-screen TV, it rises from a hidden recess in a mahogany cabinet. When he turns it off again, it sinks back out of sight so as not to destroy the rich wood-grained ambience of the roomy salon on the ex-seine boat *New Venture*. The TV is located about where the brailer would have swung over the side in the decades after Mike Sakitch took delivery of the new boat from Sam Matsumoto's shipyard at Dollarton on the north shore of Burrard Inlet in 1955. Wooden boats, especially larger wooden boats, are expensive to maintain. If the owners are handy and have the time, they can do a lot of the work themselves. Even then, tight-grain old-growth wood and marine hardware can cause costs to spiral. As licences were taken from BC's wooden boats, many found their new owners to be non-fishermen with dreams of an inexpensive yacht. All too many of these were back on the market within a couple of years, already showing the signs of neglect. Others were butchered in attempts to convert the fish hold to living space in ill-thought schemes by carpenters who had no sense of the flowing lines required for a boat. A lucky few got sensitive owners who respected the boats' integral design and fishing history. The *New Venture* had such luck.

Sam Matsumoto returned to the coast in 1949, when the Canadian government finally admitted that Japanese Canadians were not a threat to national security and allowed their return from the interior of the province. The *New Venture* was one of a series of stunning seine boats that Matsumoto turned out in rapid succession through the 1950s after which, protesting the decline in the quality of available wood, he pioneered the building of aluminum fishing boats. Other seine boats in the wooden series included the single deckers with crew bunks in the fo'c's'le: the *Eva D. II* for John Dick in 1954, the *Miss Georgina* in 1955 and the *Zodiac Light* also in 1955. Sharing the modern swept lines of these boats were a series of double decker seine boats with the wheelhouse in the top cabin, allowing space for crew bunks in the main deck cabin. These vessels included the *Ocean Star* and *Pacific Belle* in 1955, the *Silver Bounty* and *Sunnfjord* in 1957, and the *Northern Dawn* in 1961. The *Northern Dawn* was wrecked in the Queen Charlotte Islands in 2001, and the *Sunnfjord* and *Pacific Belle* were sold to Americans in 2009. Other wooden Matsumoto boats continue to grace the coastal waters, offering delight to even the casual boat watcher.

As a competitive sailboat racer and former owner of a marine supply business, Jamie VanPoele knows his boats. Before he bought the *New Venture*, he had been searching for just the right seine boat conversion. Most examples he looked at had been done poorly or had the wrong proportions. When he saw the *New Venture*, in spite of his

Jamie VanPoele, owner of the *New Venture*, is also a racing sailor.

distaste for the interior finish, he knew he had found a boat that he could live with and even live on. He could also see that it had been well cared for over the years. In the 1980s, she was owned by Mars Tarnowski. When I was working on an article about winterizing your boat, I admired the spanking white hull paint with what we called "Yugoslav green" highlights and the bright finished hardwood on the bow.

Tarnowski introduced me to a new level of care. For most of the winter lay-up he liked to keep his boat at the Campbell Avenue Fisherman's dock in the relatively warm waters of Burrard Inlet. He explained that if he kept it out in the Fraser River, the ice-cold water flowing down from the interior would transmit the chill through the keel coolers and hull fittings to the engine. This would, in turn, cause moisture to be condensed out of the air to settle in droplets on the engine and even on the engine-room ladder. So to avoid any rust resulting from condensation, he kept his boat in the harbour. But, knowing that fresh water

The *New Venture* at Vancouver's Campbell Avenue Fisherman's Terminal in the mid-1980s when Mars Tarnowski owned her.

killed any worms or growth on the hull, Tarnowski waited until the river water warmed in the spring and moved the boat to moorage on the Fraser River.

He eventually sold the boat to Bruce and Dolly Lansdowne, who owned her for several years before deciding to sell her. Their ad said she could pack 80,000 pounds of slushed salmon in her four insulated glassed fish holds. The ad explained that the "Most recent survey suggests a replacement value of $1,050,000 CDN, and a present value of $270,000 CDN." But there was a glut of good boats on the market and they were asking only "$125,000 OBO." The Lansdownes also rightly suggested that the "ex-seiner (currently packing), would make an incredible live-aboard yacht." After an English couple purchased the boat, they took her to a professional on Vancouver Island and had a well-proportioned extension added to the deckhouse. Apparently they did a bit of shakedown cruising in BC waters and then headed south. Passing through the Panama

Canal, they ventured into Caribbean waters where they spent several years before retracing their wake back to BC and putting the boat up for sale.

In 2008, VanPoele took his fine new boat on some coastal cruising. He enjoyed the experience, common to many owners of ex-fishing boats, of having old-timers approach him at various docks to tell about their experiences as crew onboard. In winter 2009, he began serious work redoing the interior of the deckhouse that extended aft over the working deck. The boat's galley and wheelhouse are in positions similar to where they were originally, but have been opened up by the removal of a separate captain's stateroom located between them. Pleasure boats don't need to travel at night as regularly as working boats so the separation of galley and wheelhouse is not so important. VanPoele has a curtain to keep the wheelhouse dark so that the operator can see forward.

The *New Venture* is a spectacular live-aboard cruising

The interior looking forward toward the original galley of the converted seiner *New Venture*.

boat for VanPoele and his partner Laura Morine. Not only does it give them a great platform from which to see the otherwise-inaccessible BC coast, but it also puts them in league with other owners of boats built by Sam Matsumoto. A benefit for other coastal cruisers is the opportunity to see a classic BC purse seiner with only slightly modified lines. For the fishermen who once worked her decks, the sight of the *New Venture* offers a sentimental reminder of good times. For young people who will not have the opportunity to work on a fish boat, it serves as a reminder of the former vitality of our fish resource.

The *Eva D. II*, built by Matsumoto in 1954 for the Dick family of Cape Mudge. Here Ivan Dick is on the bow at the celebrations for Quadra Island's first car ferry, about 1960. In 2009 she was registered to John Alexander McKenzie Jr. of Port McNeil. AUTHOR'S COLLECTION

The *Eskimo* was built in Prince Rupert by Matsumoto in 1940-41. In the late 1940s, shown here, it was fishing salmon for BC Packers. In 2009 it was registered to Coastwide Industries and owned by Kali Kaisla and his mother. COMMERCIAL ILLUSTRATORS/AUTHOR'S COLLECTION

The *Universe* was built by Matsumoto in Prince Rupert in 1942 just prior to his internment. Registered to Royal Franklin Maynard of Garden Bay in 2009. COMMERCIAL ILLUSTRATORS/AUTHOR'S COLLECTION

The 1950 Matsumoto-built *Pacific Belle* went off the Canadian registry in February 2009. *COMMERCIAL ILLUSTRATORS/AUTHOR'S COLLECTION*

The *Miss Georgina* was built by Matsumoto in 1954 when this picture was taken. In 2009 she was registered to Menzies Fishing Company Ltd. of Bowen Island, BC. *COMMERCIAL ILLUSTRATORS/AUTHOR'S COLLECTION*

The *Silver Bounty* was built by Matsumoto in 1957. Even after World War II when he moved to Vancouver, many of Sam Matsumoto's customers were from Prince Rupert. COMMERCIAL ILLUSTRATORS/AUTHOR'S COLLECTION

The *Pacific Pearl,* built by Matsumoto in 1956, and registered to James Henry Robinson of Hartley Bay in 2009. COMMERCIAL ILLUSTRATORS/AUTHOR'S COLLECTION

4
Marsons
What to Do When the Licence Is Gone

The rhythmic "thump thump" of a caulking hammer echoes across the nearly deserted shipyard. It is Sunday afternoon and most of the regular workers are taking the day off. The sound is coming from plastic skirting above, which shows the bulwarks and deckhouse-with-dodger of a good-looking wooden seiner. Wielding the weathered caulking hammer is Steve Martinolich, whose father Aldwin built the boat eight years before Steve was born.

The Martinolich family built the three *Mar* boats—*Mar-Lady*, *Marsons* and *Mar-brothers*—with Mario Tarabochia at Port Guichon just downriver from Ladner. The first two, built in 1952, share the distinction of being amongst the first drum seiners on the coast. Using a large drum mounted in a well on the afterdeck and turned through a truck transmission with power from the main engine, they were a costly experiment. But the family stuck with

The *Mar-brothers* in winter moorage at Ladner in 1983 with the *Klemtu* astern.

the innovation even as the Puretic power block was introduced for the more traditional table seines. As with all
major technological innovations, the drum seine took
some years to perfect. The nets had to be hung or made
differently. In the early days when they were still made
of cotton with hemp lines, nets stayed wet and rotted on
the drums. A decade after these experimental boats, the
majority of BC seiners were using the drum; another five
years and it was virtually universal. The technology made
setting and recovering the net so much faster and easier
that Alaska banned drum seines (though they continue to
be permitted in Puget Sound).

When I interrupted Steve Martinolich in his caulking work to ask about the *Marsons*, he explained that he
had sold the salmon licence and now kept the boat for
sentimental reasons. At 61 by 14 feet overall, the boat is
a big undertaking that causes his wife to sometimes question the sense of it all. Martinolich points out a plank just
at the water line. "I won't be changing that plank," he
smiles. "It has a history. It is flat cut so should have rotted by now and I always used to tell dad to change it, but
he wouldn't. So now I will just have to keep it." Born in
1960, Steve was allowed to spend a half-summer on the
Marsons when he was eleven but worked his first season
on the boat as a sixteen-year-old in 1976. As is the cus-
tom, he began his fishing career as the skiff man rowing
the beach man ashore, where he would tie up one end of
the net while the seine boat towed the other in a large
semi-circle to trap the salmon. Or, in the event of a drift
set, the skiff man and his partner would simply hold the
end of the net while the seine boat circled back to close
the two ends. Putting the beach man up in heavy seas can
be quite demanding of the skiff man's strength, skill and
timing. "There is one place at McInnes where you ride the

Steve Martinolich recaulking the *Marsons* at Shelter Island, March 2009.

swell up over one rock to let the beach man off on a larger rock. If you don't get back out quickly you can end up sideways between the two rocks," Steve recalled. "It never happened to me but I heard of it."

In 1986, following the regular progression with no exceptions for the skipper's son, Steve was promoted to beach man. The position demands particular skills in wrapping a tie-up peg or putting in an elaborate slipknot that could be released while there were still several tons of pull on the line from the boat's diesel engine and the tide. It was a job with just enough danger to give pride to a young man still under thirty. The next progression for Steve was to engineer in 1989. By this time, the *Marsons* had been repowered from a Gray Marine to a 270 hp GM "Jimmy" 8-71. For a man destined to own a boat, a stint as engineer is good training in the long-term care of a vessel. By the 1980s the original gear drive drum in a well was long gone and had been replaced by a hydraulically turned drum. "The 8-71 was a bit underpowered for that and the boat only makes 8.8 knots, but my dad was tight with the money."

In 1994, Aldwin turned seventy-two and had decided that Steve would take the boat out as skipper that year. As it happened, Aldwin died before the salmon season. "I don't think that he could have stayed home and watched the boat go out," says Steve, knowing himself how hard it is for a fisherman to stand on the dock watching the fleet leave for the grounds. Steve had been born to the fishery, he was raised in the fishery and now he took his place as skipper of the family boat. He had fished the whole coast from San Juan in the south to Barkley Sound on the west coast of Vancouver Island and Johnstone Strait between Vancouver Island and the mainland. He had fished the mainland inlets up through the central coast and out through Wales Channel to Drukers and McInnes Island. He had fished one of his dad's favourite spots in Masset

The *Marsons* doing what she was built for. *COURTESY OF MARTINOLICH FAMILY*

Inlet on the north side of Haida Gwaii. In fact Aldwin, like his brothers, was given to fishing every little opening no matter how out of the way it might seem. Now his son took the boat that Aldwin had built and went out to the places that Aldwin's father Matt and his grandfather Vananzio had fished. Vananzio had emigrated from the Adriatic island of Losinj, first to Puget Sound, and then up to Ladner just after the end of the nineteenth century. Steve became the fourth generation of his family to fish British Columbian waters. As always, good-paying jobs were available for relatives and others to crew the *Marsons*. But storm clouds were already on the horizon.

The fishery had been a limited entry for two decades. With the fleet apparently limited in size, some fishermen and businessmen had taken advantage of a provision that allowed several small gillnet licences to be combined to build a new seiner, so that the actual number of seiners had increased significantly with limited entry. Others had replaced narrow old wooden boats with beamy modern steel and aluminum boats that stayed within the law by maintaining the registered length of the licence while adding beam to allow for huge increases in carrying capacity, and adding power to allow for the use of heavy gear that permitted fishing in strong tides. "Fishing Wales Channel with its strong tides could be a real nightmare with only 230 hp like the *Marsons* has," Steve noted. "You had to be way ahead of your beach end in order to make the turn back in to pick up your end. The new boats with twice the horsepower and heavy gear could tow the net up against the tide."

So, regulations intended to limit the fleets' fishing power served only to increase it. Competition was heightened and actual fishing days were reduced. By the 1990s, Steve could see that fishing was not going to be a viable future for a fifth generation. But they say the fishing fever gets in your blood. Steve had good winter work repairing outboard motors and as a heavy-duty mechanic, but still he stayed with the fishery. "I never took my son out as crew on the boat because I could see which way it was going." There were good times, like the set in Wales Channel when "We filled the *Hesquiat*. Everyone else had left the area but we stayed on and I could see huge black schools of fish going into the net. When we dried up the net, the pinks were spilling over the cork line. I saw the packer *Hesquiat* nearby so called him over and we just brailed the fish right onto him. We put 90,000 pounds of pinks on but we were required to let the dogs go. We probably dumped about 25,000 pounds overboard and most would probably have died since they had been too long in the net."

That triumph in the early 2000s was tempered somewhat by having to waste the dog salmon, but it was indicative of the bittersweet vocation that salmon fishing had become. In 2003, when his uncles wanted to sell their half of the *Marsons'* licence, Steve knew that he could not justify the nearly $200,000 to buy them out. The licence was sold and Steve was left with a boat that carried only a D licence permitting it to carry or pack fish others had caught. He leased the boat to a friend who used it for a herring gillnet crew to live aboard and to pack the fish they caught with the herring punts. But this arrangement never covered the real cost of maintaining a boat of the *Marsons'* size.

In the summer of 2008, Steve went on a six-week voyage driven by the urge to be up coast and to show his sons, nineteen-year-old Matt and eleven-year-old Sam, the signature places of his life. Stopping in at old fishing spots and special places his father had shown him, Steve told the stories and showed the sights as few non-fishermen

Its salmon license sold off, the *Marsons* is still kept up by Steve Martinolich for recreational use. *COURTESY OF MARTINOLICH FAMILY*

have seen them. Some northbound highlights included Texada Island and Chatham Point, then into Rivers Inlet to visit the old canneries. After visiting Quay River, where daffodils still grow in an old homesite and the beach is rich with clams, they spent a few days in the ruins of the cannery town at Namu. The mayor of Namu, a kind of caretaker, was there to greet them and tell them to just obey the signs that turned out to be nearly everywhere and all said the same thing: *DON'T*. But they did enough to see what was once the recreation centre (with pool hall and bowling alley) kept from falling down the slope by a single stump on which it crouched. Then they cruised on to Bella Bella and across the way to go ashore for a meal at Shearwater, where a picture of the *Marsons* in her fishing days hangs on the wall.

At this point, Steve was entering his most familiar

territory. They cruised past Ivory Island Lightstation and anchored around the old haunts up inside Mathieson Channel. "I showed the kids a pink salmon stream in Salmon Bay and we went through Sheep Pass and caught halibut in Windy Bay," Steve rolls off the names like a chant. Steve was able to show the boys Mussel Inlet, Poison Bay at the head of Mathieson Channel and Carter Bay near the head of Finlayson Channel. He told of how three of Capt. Cook's crew ate poisoned mussels that led to the death of one named Carter, who was buried at Carter Bay. Also at Carter Bay, he showed them the bow of the steamer *Ohio*, which sank after hitting Ohio Rock in 1909. For Steve every rock, point and bay on the coast has a story.

Moving up coast, they anchored in Green Inlet and then Khutze Inlet. "I took the boys across to Canoona River, which Dad always called Indian River, because I wanted them to see a Kermode bear, but we had no luck," he recalled. "We used to take a week to come back south after fishing the north and we would stop in at all these places. My dad once took me up Indian River and we rowed a ten-foot skiff up the first lake to a narrows with a sandbar covered in clamshells, so we knew there had been an Indian village there because we were a long way from the sea." They ventured on to the ghost cannery at Butedale that, like Namu, is being reclaimed, along with its hydro plant, by the coastal rain and rich plant growth. From Butedale they went to the hot springs in Bishops Bay: "You practically have to take a number now," Steve lamented. "It used to be just fishermen but now there are tourists."

In a way, Steve and the boys had themselves become tourists. But only Steve could tell the story of his dad's cook, the Nova Scotian Joe Richards, and how their grandfather had scattered Joe's ashes on the sea off this rocky bit of coast. "I took them out to catch a humpback salmon with a rod and reel at McInnes so that they would be at least the fourth generation of Martinolich to catch a pink salmon there." Southbound, the three travelled to Port John where the boys saw the rock walls below the high tide that were traditional Native fish traps for the salmon that come to the creek there. Farther south, they stopped at Alert Bay. No sooner had they tied up than a slightly inebriated man stepped aboard and invited himself to a cup of coffee in the galley declaring, "I am always welcome on a Martinolich boat." Steve and the boys listened as the man told stories of fishing on the *Mar* boats in the early 1950s. He then took them all uptown for ice cream. This is a common occurrence for people travelling the BC coast with a well-known vessel. A hundred or more people may have worked on such a boat over the decades and for each of those people, no matter how brief their time aboard, there are stories to be told.

Steve Martinolich's two boys enjoying the sunset on their coastal trip on-board the *Marsons*. COURTESY OF MARTINOLICH FAMILY

In the lower Johnstone Strait, the *Marsons* anchored out just above Chatham Point and Steve told his sons the derivation of the name of nearby Green Sea Bay. A seine boat called *Green Sea*, like most early boats, had a small wheelhouse on deck with the galley and living quarters below in the fo'c's'le. Apparently the skipper was down there when the boat ran aground. Not a proud moment, but one that other fishermen might well commemorate with a local name for the bay. Steve's uncle, Richard Martinolich, once told me his father Matt bought the 65-foot seine boat *Green Sea* in 1919 and sold it a few years later, perhaps after the Green Sea Bay embarrassment, and the boat sank in Barkley Sound with a load of pilchards in 1924. From the same anchorage near Chatham Point, the modern Martinolich trio took a speedboat they had on deck and explored the Yuculta Rapids and some of the mainland shore. Farther south, the crew stopped at De Courcy Island before wending their way down through Dodge Narrows and back through the Gulf Islands to their home in Ladner.

Steve's oldest son Matt is thinking of becoming a fireman but Steve looks wistfully to eleven-year-old Sam and says, "He has those big Martinolich hands and the patience that it takes to make a fisherman." At the same time, he believes that those days are probably gone for his family and if they go back as far north as Wales Channel again it may well be in a fast fibreglass cabin cruiser. They could trailer it to the end of the road at Kitimat to make the run down Douglas Channel to the old fishing grounds. Even as he says so, he returns to caulking the venerable *Marsons*, because he knows that a tourist boat is no way for a Martinolich to visit the north coast.

5

Randy Reifel

Patron of the Fleet

Randy Reifel owns some good boats but two stand out for the significance of their original owners to two generations of British Columbia fishing. Both were built to the demanding standards of men who, like Randy, were not born to fishing families but grew up in fishing communities. Both men, in their own times and places, were drawn to the life of the fisherman from the more settled ways of their fathers.

The *Sleep-Robber*, built in 1956, was neither Fred Kohse's first boat nor his last, but it is the only one of his boats still on the BC coast—and it is undoubtedly one of the coast's iconic vessels. Fred grew up around Victoria harbour in the early years of the twentieth century. His father had worked on the construction of the Empress Hotel and later operated a boat-rental service. Fred was guiding sport fishermen in 22-foot clinker-built boats by the time he was nine. One of the era's early rumrunners, Johnny Schnarr, lived in the family home so Fred got an

The resurrected *Sleep-Robber* ready for the water in September 2002.

early introduction to the challenges and opportunities of the sea. But to his disappointment, his father moved the family north in the 1920s to homestead a patch of forest at Port Hkusam just below Kelsey Bay on the east coast of Vancouver Island. One day, while cutting a big tree to let the sun shine on his dad's hayfield he watched a seiner come into the bay and take a set of 11,000 chum salmon. Two years later, in 1926, he and a brother-in-law rowed some 200 kilometres up to Smith Inlet to gillnet salmon. Kohse proved himself as a fisherman and went on to build a series of wooden and steel fishing boats to keep pace with his seemingly limitless energy and enthusiasm for the industry.

A generation after Fred Kohse rowed to Smith Inlet, another boy grew up in a BC coastal fishing community, in Alert Bay, surrounded by fishermen. He occupied what can be the most awkwardly outsider position in a small town. Byron Wright was "the minister's kid." It was the 1950s and a time of dramatic change in the BC seine fleet. The United Fishermen and Allied Workers' Union had managed to get a decent price for fish when they united the Native and non-Native fishermen in the 1938 strike. The post-war years had seen the introduction of war-time technology that provided better diesel engines and hydraulics to aide the hauling of purse seines to increase sets and catches. As a teenager in the early 1960s, Wright talked his way aboard boats like the *Twin Sisters* with Jimmy Sewid and the *Chief Takush* with Simon Beans. Don Pepper, whose dad ran the cinema in Alert Bay, recalls that his friend Byron made his first real fish money with a borrowed gillnetter. Like Fred Kohse, Wright went on to own a series of boats culminating in 1990 with the Shore-built aluminum seiner *Prosperity*. Reifel bought this boat some years before Wright died in 2008. "Everyone told me that he was a tough man to deal with but I found him fair in every way," Reifel said.

The aptly named *Prosperity* was a top producer for the late Byron Wright, who had a reputation as an aggressive fisherman.

Reifel owns eight of BC's finest purse seine boats. Collectively, they represent the dreams, aspirations and successes of dozens of fishermen. These are and were fishermen who gambled to build or buy the boats, brought family to celebrate the launchings or purchases, and drove them through storms. But most of all they carry with them stories of difficult fisheries and full holds. These are the stories that crewmen, some of whom may have spent only a season or two aboard, tell their children and grandchildren. But none of Reifel's boats are retired yachts or museum pieces. Even the wooden *Sleep-Robber* maintains a fishing licence and is ready at a day's notice to be rigged for the grounds, continuing to employ crews who set out with hope and expectation of making a living with their share of the catch.

By all accounts, Reifel follows the practice of all good vessel managers. His crews have whatever they need to catch fish with a reasonable level of comfort. This support extends to the huge, covered newsprint barges he has converted to net lofts and a well-stocked machine shop. In his shop, Lyle Whitter and Terry Steves add several more decades of know-how to the operation. Lyle worked on tuna boats in Zamboanga in the Philippines when Fred Kohse and BC Packers operated the Mar Fishing plant there in the 1970s. He worked many winters overhauling boats for Ritchie Nelson at his Queensborough Shipyard, and can tell of that legendary fleet owner and processor wading in to pitch fish on the packer *Western Express* when he had a cannery at Bristol Bay on the Naknek River in Alaska. Terry Steves spent a part of his career in the Britannia Shipyard in Steveston, where he grew up as a member of the fishing settlement's founding family. All of this knowledge and more carries on the fishing tradition on the Ladner waterfront. "I make my money in mining," said Reifel,

Randy Reifel had the galley of the *Sleep Robber* refurbished as well as the main cabin sleeping quarters. It is hard to imagine this as the same deckhouse that was flooded by a freak wave in the Gulf of Alaska over 40 years earlier.

who grew up in the Ladner area and went to school with the children of some of the coast's top fishing families. "I just expect the boats to pay for their upkeep."

"But what," I have been asked, "is Randy Reifel doing all this for?" This is not a question that Fred Kohse or Byron Wright would have had to answer. Nor would anyone who grew up in a fishing town in the 1950s and '60s. The fishing kids were the guys who came back to school in September, or a little later if it was a good dog salmon year, with the best clothes, fanciest cars and prettiest girls. More than that, they had been out there someplace away from the town. They had been on adventures that the logger's, farmer's or minister's kids could only dream of. Memories of life in the fishing community of Ladner never left Randy Reifel. "I grew up on Westham Island just across Canoe Pass from Ladner and went to school in Ladner. My friends were fishermen but I was never involved," he told me as we admired the massive new propane cook stove in the redone galley of the *Sleep-Robber*.

"But why have you done all this?" I asked.

"I enjoy the boats and the people," he explained. "It is a cliché but they are the salt of the earth." Some wealthy men donate to a museum, perhaps even a maritime museum where old fishing boats are preserved at great cost and with only a modicum of success. But Randy, taking advantage of Canadian fisheries regulations that do not require the owner to also be operator of the fishing boat, has instead built a living, functioning, self-sustaining operation that acknowledges the best of the past while providing a means for survival into the foreseeable future of an honest-to-goodness British Columbia fishing community.

All but one of Reifel's eight boats carry two seine licences, allowing them to fish both north and south areas. He has also purchased pilchard and herring licences, as well as dogfish and halibut quota for longline

Lyle Whitter in the engine room of the restored *Sleep-Robber*.

fisheries. The fleet fishes for Bella Coola Fisheries, located upriver from Ladner. Jack Groven, one of the partners in Bella Coola Fisheries, uses his market knowledge to coordinate the fishing effort of Reifel's fleet. Like most good leaders, Randy is fascinated by the details of his operation but doesn't hesitate to delegate to those with more specific knowledge. He is also quick to give recognition where it is due. When showing a guest the engine room of the *Sleep-Robber*, he insists on Lyle Whitter coming along to describe how the six-cylinder 425-hp D353 Cat engine, drowned in a sinking, was resurrected. "After they raised the boat they flushed out the engine and started it up," Whitter explained. "But they should have run it long enough to fully dry out all the parts. When I got to it the pistons were so seized up that I had to drive them out of the cylinders with a block of wood and a sledgehammer."

That was just the start of the rebuilding of the engine and engine room. When the *Sleep-Robber* was designed by noted naval architect William Garden and built at Mercer's Star Shipyard in New Westminster, the use of hydraulics was just being introduced to the industry. Over the decades, hydraulic lines had been run from the engine room to any place on the boat that human muscle or shaft-driven power could be replaced by flexible hydraulic power. Where booms were once raised by wrapping the topping lift line around a mechanical winch capstan, they are now lifted by a small hydraulic winch at the base of the mast. When the *Sleep-Robber* was built, herring were brought aboard with a winch-operated brailer. Later a submersible hydraulic fish pump replaced this method. Each of these modifications required the addition of another hydraulic line. "It looked like an Italian spaghetti farm down here," Lyle said, "but we took all that out and replaced it with swept or properly bent piping with high quality tig welds."

It is affirming that there are individuals such as Randy Reifel, along with those who work with them, who can appreciate the classics of the coast. Under current Canadian regulations that allow non-fishermen to purchase fishing licences or privileges, the role people like Reifel play is important in financing the continuation of commercial fishing. However, if regulations were modified to limit control of fishing licences to fishermen, the price of a licence would be significantly reduced. This would be a very different reality on the coast where those who crew the boats would be able to own them as well. In the meantime, the crews that fish for Randy benefit from his role as patron of the fleet.

Sleep-Robber

Not Ready to Be Counted Out

It's not unusual for engine rooms in fifty-year-old wooden fishing boats to look and smell like a medieval dungeon. If the boat has been sunk and oil has coated everything, it will look even worse. That is how Lyle Whitter had found what is now the bright and highly organized engine room of the *Sleep-Robber*. The story of how the boat got to that condition, and how it was brought back from a fate of being broken up and put into dumpsters like the *Kamchatka*, is a testament to wooden fish-boat rebuilder extraordinaire Brad Scott, along with other fine tradesmen and the support of boat connoisseur Randy Reifel. When Fred Kohse died in July 2001, the obituaries listed the many boats he built and sailed to the Pacific and Atlantic Oceans' farthest fishing grounds. Around galley tables up and down the coast, old-timers recounted the tale of Fred's famous escape with the *Sleep-Robber* from a rogue wave in the Gulf of Alaska. But few realized that the eulogizing of Fred's life was only the foreshadowing of the *Sleep-Robber's* next brush with death.

The 66-by-20-foot *Sleep-Robber,* which Kohse owned until 1971, was the boat in which he explored the tough Bering Sea halibut grounds and contributed a number of stories to the coastal classics. The boat raised some eyebrows when it was launched. Impressed with the large beam the Americans were building into their 58-foot Alaskan limit seiners, Kohse reasoned that the extra beam, for its day, would give him bigger payloads on the long halibut trips to the Gulf of Alaska and the Bering Sea. Kohse named his new boat *Sleep-Robber,* not simply for the long hours he expected to work her on the grounds, but in admiration of Mao Zedong's guerrilla fighters who had earned the name by stealing guns from sleeping Nationalist soldiers. The boat worked well for Kohse as he travelled the open north Pacific waters with full loads. In the spring of 1964, Kohse was returning heavy with a load of halibut across the Gulf of Alaska from a Bering Sea trip:

> I think one big wave came down on us like that. We were running on a slight angle, before the swell. There was almost no wind—I don't think it was blowing any more than 15. We were going, not full speed, but quite nicely. Rick Johanson, who later had the *Bravado* and was lost, was on the wheel. The boat had been on the mike (auto pilot) but she had been sheering off a bit and Rick had taken over. I was lying down. All of a sudden there was this horrendous crash. I thought for a minute it was an airplane landed on us. There was a breaking of lumber and smashing. I couldn't figure what the heck was going on. I looked down and there was nothing but wreckage all around. The swell

had come down on the starboard side. It smashed twenty feet out of the side of the house. The heavy plywood was all broken up in little chunks. It lifted the cap off the rail and sprung some of the planking above the water line. We lost our steering when the hydraulic lines were torn out with the cabin wall. In the engine room the engine was half under water but the breathers were out so it kept running. We had one of those automatic inflating life rafts on top of the cabin, about fourteen feet up; it was torn off and inflated, but it was still attached to the boat by a line.

When the wave hit, Charlie Madden, Phillips and, I think, Ed Bowen were lying down in the living quarters. Jacob Jacobson, the cook, and Jimmy Riley were in the galley. The galley windows were covered with plywood storm windows but the wave broke out through the galley door on the back of the cabin. The force of the water broke the lock on the door. There were glasses in a rack high on the galley wall. After, we found these to be full of water. The water hit with such force that the four-inch aluminum table support was bent right over. When the water went out of the galley, Jacob was left with two cracked vertebrae and Riley with his leg hurt. No one was hurt in the crew's quarters, but all of their clothes and bedding were washed down into the engine room. The clothing kept plugging the intakes on the pumps so I had to keep diving down to unclog them. Luckily the auxiliary was located high enough so that we had this to support the main engine pump. We lost electrical power for a while when the lines were torn out, but we managed to get it going again. We threw the clutch

out and lay there drifting while we got to work on
the boat. We fixed the steering and spent ten hours
pumping out the water. It took a few hours before
we knew that we were gaining on the water and
that we would be okay.

When the *Sleep-Robber* limped into Vancouver some time
later, the newspaper featured a photo of Fred Kohse standing in the gaping hole in the side of the cabin looking
justifiably proud of himself and his good boat.

After Kohse sold the boat in 1971 it passed through a
number of hands and suffered other calamities, including a fire that nearly destroyed her. In 1992 Barry Curic
bought her. There was still some money to be made in
the BC fishing industry, but Curic fished the boat through
some increasingly tough times. By 2001, the changes had
begun to bring serious declines in fishing time and catches. In August of that year, Curic was glad to get some good
pink fishing on the north coast. He later told how good it
felt travelling into Prince Rupert on the night of August
16:

> We had a good jag of fish onboard, about 60,000
> pounds of pinks. I was on the wheel and could
> hear the guys all laughing while they ate their late
> supper. At about 11:30 I was relieved (on the wheel
> watch) and went down to have some supper and a
> short rest. With two men on the wheel I felt secure,
> but then came that terrible crash. We were nearly
> into the harbour when we ran up on a pinnacle at
> Lima Point on Digby Island. It was just after the
> turn of a 21-foot-high water. I called on the phone
> to see if anyone was around but we were the last
> boat in so there was no one for a tow off. We tried

Dropping tides have rolled the *Sleep-Robber* onto her side. *COURTESY OF DAVE LEWIS*

backing off but she wouldn't budge. I told the guys to tape up all the vents and seal the portholes and doors, then to get their stuff into the dead skiff. Finally I got in. Then I jumped back on the boat to go down and shut the engine off to minimize the damage. While I was in the engine room she started to go over. I had just killed the engine so I ran for the ladder but it was hard to get to it as she was turning. Then she stopped and I got off. About five minutes later she went right over. It was as though she was giving me that bit of time.

The boat lay over on her side at low water, but Curic still had hopes of getting back his pride and joy. With the boat on her side, the waters flooded the hull when the tide came back in. Fortunately they were not in an area of heavy seas that would grind a boat to bits in a matter of hours. Curic arranged for a floating crane to strap and lift the boat, but then problems in the salvage operation led to the boat being dropped and sustaining more broken ribs on her port side. The mast was torn out along with part of the bulwarks. Curic found it incredibly hard to watch. "I thought I was pretty tough," he says, "but when I phoned my wife I just broke down."

Curic is quick to point out that many people willingly offered help. Kevin Smith, who had just rigged his boat, the *Western Investor*, delayed leaving on a trawl trip. "Kevin and his crew came over and helped pull my net off the boat. A former deckhand came down and helped me salvage stuff out of the wet cabin and Canfisco manager Chris Cue was there with help at every turn. Stuff like that is really appreciated." When I spoke with Barry Curic a month after the accident, he didn't think there was much hope of rebuilding her. His insurance company agreed and he had been looking at some of the other wooden boats that were for sale at the time. But his heart was still with the *Sleep-Robber*. "It's not like it used to be and you couldn't build a boat like the *Sleep-Robber* even if you wanted to," he said.

But the noble *Sleep-Robber* was not ready to be counted out. Brad Scott, who had restored several large wooden boats, bought the hulk from the insurance company and took it out of the Fraser River at Shelter Island Marine. Scott's initial plan had been to salvage all the machinery other than the seawater-seized engine, and scrap the boat. Before he could do this, Randy Reifel offered to buy the boat if Brad would oversee the rebuild. Reifel's confidence in Brad's ability was justified by a number of big wooden boats that Brad has brought back from the brink, especially the 103-foot halibut schooner, *Tyee*, built in Seattle around 1914. Scott brought the *Tyee* back from a certifiable derelict in the 1980s to become one of the finest fish packers in the BC fleet. More recently, he owned and operated the classic *Phyllis Cormack*. When she was wrecked on a beach, he bought the 1938 pilchard seiner *Midnight Sun* and packed salmon out of Alaska with her before selling her and carrying out an extensive and luxurious yacht conversion for the new owner. In 2009 he owned the *Good Partner*, on which he had rebuilt the deckhouse after a fire.

Even for Brad and his crew of experienced and skilled shipwrights, the *Sleep-Robber* represented a challenge. Brad hired Lyle Whitter to rebuild the engine and mechanical components. But they were the least of Brad's worries; mechanical challenges are quantifiable and replacement parts are available. In contrast, the wooden parts of a boat are not interchangeable. They are the result of careful design and handcrafting. Damage resulting from the salvage was readily apparent on the port side of the boat, where ten feet of the hull was staved in a good three inches. The fibreglass and foam from one of the four fish tanks had to be removed so the seven ribs could be replaced. What

With kelp in the rigging and water in her hull. COURTESY OF DAVE LEWIS

The *Sleep-Robber* showing her bottom on the rocks near Prince Rupert.
COURTESY OF DAVE LEWIS

was not expected was the amount of rot they found in the other frames. As work progressed, it became apparent that a total of thirty frames per side would need to be replaced. From the forward engine room bulkhead aft to the lazaret, all of the frames were sistered with second frames down beside the original. Some of these were part of the way down, with a number of new frames being driven down right to the keel. Among BC seine boats, the *Sleep-Robber* and her sister ship the *Blue Pacific #1* are unusual for having flat transom sterns rather than more rounded timbered sterns. The transom was in good shape, though some of the frames had cracked at the bend of the chine.

Each side of the hull has about eighteen courses of planks, with an average of three planks to the strake. About 40 percent of these were replaced and all of the 2.5-inch planks were refastened. The nearest source for ship's spikes for the job was in Massachusetts. When Scott found it was going to cost $1 CDN for each 4.5-inch nail landed in Vancouver, he decided to use galvanized 3/8-by-4-inch lag

Brad Scott and his crew saved the *Sleep-Robber* from the wreckers.

screws instead, as he could get those for 45 cents apiece. In the end, they put four thousand spikes and ten thousand lag screws into the hull. The colour and grain of the new planks on the *Sleep-Robber's* forty-six-year-old hull stood in stark contrast to their aging neighbours.

Deregulation and globalization make it increasingly difficult to find adequate wood supply for the maintenance of fine vessels like the *Sleep-Robber*. Scott has much of the wood for his boats quarter-sawn in BC at Image Works in Coombs, and laments that logs from the Crown lands belonging to the people of BC are far too often loaded onto freighters and shipped to the highest bidders on the world market. When I visited the shipyard during the *Sleep-Robber* rebuild, shipwright Pierre Pereira pointed out the bright new golden of the edge-cut, old-growth fir planks. At thirty lines to the inch, the tight growth rings form a tight barrier to the water. But more importantly, Pereira pointed out, "When these planks swell with water, they expand sideways and compress the caulking, while the flat grain planks take in water and swell outward." This is the kind of knowledge that is second nature to a good wood shipwright, and the kind of knowledge that is essential to the mammoth task of bringing the classic seiner *Sleep-Robber* back from near death.

7
Dan Nomura
Still in the Business

Dan Nomura's resumé credits him with a Master of Science in zoology from the University of BC, but his pride and his heart are in those things that his father taught him on the stern of a salmon troller. One of those things was the understanding that in BC, "salmon fishing goes beyond making a living, it is a way of life." Nomura explained this to me one day shortly after the 2009 herring season. I was visiting him in the bright red Canadian Fishing Company home plant at the foot of Vancouver's Gore Avenue, his office overlooking part of the seine fleet moored outside. Dan's father Junichi took his fourteen-year-old son out as a deckhand on a combination troller-gillnetter. Later, Dan worked on his uncle's packers and in 1976 became skipper of the packer *New Whitecliff*. Dan's dad was born in Canada but was of a generation that was still sent back to Japan for their schooling.

For his master's thesis, Dan did extensive research on the effects of weak electric fields in catching salmon. It

Dan Nomura, the president of the Canadian Fishing Company, in his offices at the foot of Gore Avenue in downtown Vancouver.

had long been known by commercial salmon trollers that the electronic fields around their boats could have an effect on their catching power. Dan worked with his dad and other trollers to find that a neutral or slightly positive charge could help the average fisherman attract a few more fish to his gear. It was an interesting affirmation of what fishermen had been saying for years. But more importantly it gave Dan, as a young adult, an opportunity to do advanced study on his dad's philosophy as a fisherman. "Dad made his money trolling," Dan recalled. "He liked trolling as a gentleman's game with boats showing respect for each other. But when the fall came, he put on his chum salmon gillnet just because the others were doing it and he didn't want to miss out on anything."

"My father thought of fishing as a hard living," Dan explained. "His generation didn't feel pride in being a fisherman. They didn't give themselves enough credit for their innovation and determination. So they worked to make it possible for their children to have the education that they believed would lead to a better life as professionals." While this is the route Dan followed, he has never really left the fishing his father taught him to respect. On his graduation from UBC, he took a position as a research biologist with the Pacific Biological Station in Nanaimo. But his upbringing in the commercial fishing industry drew him back and he became a manager with the Prince Rupert Fishermen's Co-operative. In 1982, he joined the Canadian Fishing Company as manager of quality control and technical services.

It was a difficult time in the fishery, with skyrocketing interest rates resulting in banks seizing boats they had no idea what to do with. Fishermen with decades of service were losing their boats and their houses. Things weren't any better on the corporate side. Up-and-coming

Seiners at the Canfisco dock at the foot of Vancouver's Gore Avenue include the *George McKay* and *Salli J. Rogers* built by Matsumoto in 1971, the *Western Ranger*, built by Shore Builders in 1977, and the *Cape Spruce*, built by Shore in 1987.

entrepreneur Jimmy Pattison bought the Canadian Fishing Company when the US parent New England Fish, faced with bankruptcy, was obliged to dispose of it as an asset. "Mr. Pattison buys a company when the price is right, he can see potential and he likes its management team," Dan said of the purchase, which happened shortly after he joined the company. The rapid changes in fisheries management combined with a general sense of uncertainty may have also led to the Weston family's decision to sell BC Packers, which had earlier taken over Nelson Brothers and Nelbro, their Alaskan subsidiary. While BC Packers cashed in their extensive Steveston real estate, Pattison saw an opportunity for consolidation and bought their processing facilities, fleet and licences. At the same time, the federal government was in the midst of a program to purchase and retire salmon licences, which enhanced both the catch potential and the value of licences remaining in the fishery.

But this was not a time when new fishing grounds were being discovered, nor were runs getting larger. Dan remembered it as a time when processors had to become competitive in a global market by being more efficient, maintaining premium quality, and diversifying by adding value to their products. In 1998, Dan became Senior Vice-President of Canfisco and their Alaskan operation, Kanaway Seafoods, which eventually became the large Alaskan subsidiary Alaska General Seafoods. Dan smiled as he recalled the traditional separation of processors and fishermen that had his dad saying, "If I had known that you were going to become a manager in a fishing company I would have made you a fisherman."

But it was his fishing experience that gave Dan the practical knowledge to be a manager. Recalling his first time in command of a salmon packer in the mid-1970s, he told me,

The Canfisco packer *Koprino*, built by Hoffar & Beeching in 1926 and rebuilt in 1951, is one of BC's classic wooden boats. In this picture it is collecting gillnet salmon and flying the red and white Canfisco company flags.

"After my uncle, Sid Teraguchi, said that I was ready to take his boat up to Rivers Inlet, I hired a friend, Terry Shiyoji, and we went down the Fraser River. When we came into the Gulf of Georgia there was quite a bit of wind and tide off the light ship. We hadn't lashed down the scales and other stuff on deck was all sliding around. We stopped there and were trying to get everything in order when Ben Robinson of the Kitasoo First Nation came along and asked where we were going. When we told him that we were heading for Rivers Inlet he noted that we weren't off to a very good start. But he also invited us to travel with him." It was the beginning of one of those relationships that endures. Dan continues to work on committees with Robinson, including a scallop aquaculture initiative.

The recognition of hereditary rights of First Nations in the salmon fishery and some other inshore fisheries will

Canfisco's classic steel seiner *Royal City* (Star Shipyards 1963) runs against the flood tide in Johnstone Strait with Aubrey Roberts at the helm.

The venerable packer *Hesquiat*, built in 1944 by Newcastle Shipbuilding in Nanaimo and owned by Jim Pattison Enterprises in 2010.

shape the participation of Canfisco in the future of the salmon fishery. "I expect that we will be selling some of our boats to First Nations who will use them to fish their licences. In return we will look for a long-term fishing agreement with the owner with financial and infrastructure support." At the same time, some of Canfisco's smaller aluminum seiners will be deployed to Alaska. As Alaska has an owner-operator provision for their salmon permits (licences), the boats will be sold to fishermen along with a long-term fishing agreement between the fisherman and Alaska General Seafoods. This concept is similar to what Dan envisions could take place for some BC First Nations and Canfisco when a salmon quota system is in place.

Over the past twenty-five years, Canfisco has grown through acquisitions such as BC Packers, as well as strategic arrangements like Pattison's partnership with Ocean Fisheries. This positioning has allowed for central access to a major portion of the salmon and herring production in BC, and also demonstrated a long-term commitment to the health of the fishery and its commercial business. Pattison named Dan Nomura to the position of company president in 2006. Dan brought with him the responsibilities his immigrant grandfathers and his Canadian-born father passed on to him, a tradition of community service and continuity. He takes pride in Canfisco's preservation of the 1937-built pilchard seiner *Snow Prince* as a boat to take out its customers and staff. Together with another salmon troller's son, Paul Kariya, Dan was instrumental in working with the Nikkei Fishermen's History Book Committee to produce two excellent histories that document the Japanese-Canadian contribution to commercial fisheries and to British Columbia. These projects are precisely the sort of involvement that serves Canfisco well in defining its respectful position in the fishing industry.

"Sisu"

Gus Jacobson and the *Evas* of Finn Slough

When the federal government increased the property rights of commercial fishermen so they could lease their licences to others, not everyone took the bribe. Some who wanted to leave a fishery took the high road when they sold their licence. "When I sold my herring licence, I stipulated to the broker that it must go to another fisherman, not one of the corporations," Gus Jacobson told me in 1999. Turning sixty that year, Gus had the trim build of a man who has enjoyed life and worked hard at it. On the evening I visited him at his family home on the south side of Lulu Island in the Fraser River estuary, he had spent the day helping friends hang herring gillnets. "Now I will just pack fish for other herring fishermen," he explained. "I'm in pretty good shape but last year my son and I shook 16 tons of herring in a two-hour opening and then we did 27 tons in an eight- or nine-hour opening.

The classic Fraser River gillnetter *Eva* at home in Finn Slough.

With those short openings you don't have time to get in shape."

Gus's Finnish "sisu" (tenacious) quality was passed down from his immigrant grandfather Mikko Hihnala. Having changed his name to Jacobson during a stint working in a coal mine near Astoria, Oregon, Mikko brought "sisu" north to Canada and the Fraser River in 1892. By the 1920s he had been joined by other Finns to form a sizable fishing community living mostly on government land outside the Fraser River dike. Partially to avoid paying land taxes and eroding banks, they moved in the late 1920s a short distance upriver and began building scow houses and stilted houses on a little slough along the river formed by Gilmore Island. They continued to squat on the land and the adjoining waters with small houses, netsheds and floats in what came to be known as Finn Slough. Over the years, some bought land nearby on what came to be Finn Road. By the 1970s, most of the people had larger homes back from the slough while they continued to keep their boats and nets in the slough. The original little houses have since been rented or sold to a new group of people—many of whom take creative inspiration from the quiet haven of the slough on the edge of what is becoming increasingly an area of urban sprawl.

Gus recalled that in 1939, the year he was born, "Dad paid for a new house, a new boat and a new car. The car was the most expensive but the two-storey house is still there on Finn Road." He went on to tell of buying his first boat, the *Ina*, from an uncle in 1953 when he was fourteen years old. "I paid $600 for the boat with three nets and I got a netshed with a little attached house along with it." Fishing only the Fraser River, he immediately made about $1,000 in a salmon season that stretched for several months and allowed four and five days of fishing per week.

> For many years I made a living here. My dad was a spring fisherman and I followed him. You had to fish in among the snags on the flats. The fish came upriver on the flood tide. But the tide didn't slacken on the surface until two and a half hours after the beginning of the flood. The level at Finn Slough would have already raised a half to a foot by that time.

Gus explained the intricacies of river tides that a fisherman must know. Apparently the heavier saltwater pushes up in a wedge under the surface water that continues to flow downriver even though the tide is pushing upriver. "When a southeast wind is blowing down the river it will actually build the tide in the river while a westerly should blow the water into the river but the tide will actually be a foot lower with that wind."

Gus made a good living from the salmon fishery. He married and raised a family along the river. For many years, Gus had kept, and on occasion fished, his uncle Henry Jacobson's classic 29.5-foot gillnet boat, the *Eva*. In his uncle's days, a frugal family could buy a boat and pay for it with a good year's fishing. This was possible because the boats were less grandiose and licence fees were nominal, but also because the season started with early spring (chinook) salmon fished in the river in late February and March. It was a time when fishing was left pretty much to fishermen, with the processors sticking to their canneries. But it has been decades since there was a commercial fishery on those early springs. When the federal Department of Fisheries implemented a limited-entry strategy in the

The *Eva* on the Fraser River about 1988 when she still had a licence.

late 1960s called the Davis Plan, after Minister of Fisheries Jack Davis, Gus supported the concept because he saw that the growing fleet needed to be controlled. The plan included categories for low-production and higher-production boats in an attempt to phase out part-time fishermen and leave the fish for the people for whom the industry was central to their living. Gus's gillnet boat qualified for the A-Licence category as he was a good producer. At first he believed that secured his position in the industry, "But then they kept stripping [areas and other privileges] away from it."

A part of the plan, which licensed boats rather than fishermen, allowed for licences on several small boats such as gillnetters to be combined and retired to allow for the building of a larger seine boat. Between 1969 and 1975, the number of seine boats increased from 369 to 483, as did the cost of their individual transferable licences. In one case, a non-Native fisherman travelled to upcoast

First Nations villages, buying up small gillnet boats and consolidating the licences to make a seine licence. He then invited doctors and others looking for tax deductions to invest in new boats for which he provided the licences. The plan eventually went into bankruptcy but the boats remained in the fleet. The changes brought by the Davis Plan have been gradual. The requirement that boats choose to troll or gillnet limited the range of opportunities for an individual fisherman while apparently strengthening that of fishermen committed to a single gear type. The absence of a provision in the licence plans for the owner of the boat to also fish it, as was done in Alaska, had little effect on the gillnet fleet. The large fishing companies could benefit more from the production of purse seine boats and so their money tended to concentrate in that fleet.

By the 1990s, river openings could be measured in hours. Gus had let a friend fish the *Eva* while he and his wife Pat worked their modern 34-foot fibreglass *Huki Lau I* farther north in Smith and Rivers Inlets. The Mifflin Plan, named colloquially for another Minister of Fisheries, divided the coast into three gillnet areas in 1996. Each area required a separate licence, so Gus and Pat stacked the *Eva*'s licence on the *Huki Lau I* to allow them to continue fishing in both the north and central coast areas. Gus couldn't part with the *Eva* or the fishing heritage she represented. He considered a couple of commercial fishing museums. "But a boat has to be used or it dies," he said. "And the boat has been a part of the slough since Kishi built her for my uncle sixty years ago." In the 1970s Jim Kishi, the son of the *Eva*'s builder, replaced the frames in the *Eva*'s fishhold with the same care and craftsmanship with which his father had originally installed them.

It was on Gilmore Island in Finn Slough, with its little

cedar-shake cabins balanced on piles just above the water at high tide, that he would find the resolution for his dilemma. Twenty-five years earlier, Al Mason had moved into and fixed up a 350-square-foot cabin on the island. He did a lot of different things but mostly he worked with wood, and he fell in love with Gus's *Eva*. Recalling the continuing infatuation, he said, "I always coveted her but I knew Gus would never part with her." In 1983, when Al found himself between jobs, he began making a 1:12 scale model of the little boat. After three months of long days, he had his own *Eva* complete with the distinctive *kerput* sound of her Easthope engine, tire fenders, and a working gillnet drum from which he could remotely set her net. Three years later, the little boat was a hit at Expo 86 in Vancouver where Al demonstrated her daily for several months. Then she came home to rest on a ledge near other wooden models of boats, trains and trucks. "I don't think of her as a model," explained Al, who has spent the

The *Eva* was built in Steveston by the legendary Kishi shipyard. PHOTO BY AL MASON

Al Mason built a scale model of the *Eva* then traded it for the actual boat. Here he works on the restoration of a 110-foot Miki-class tug.

last several years rebuilding a client's 110-foot Miki tug. "I built her as a little boat."

Talking with Mason one day, Gus made a startling offer. "Why don't we trade boats?" he asked. It wasn't the first time Al had traded a model for the real thing, having once swapped a wooden truck for a pickup, but it was definitely the finest trade he had ever been offered. When Al answered an immediate "yes," Gus, recognizing how important that model was to Al, suggested that he take time and think it over. But it wasn't long before the two men had agreed to part with their treasures.

In 2009, I visited Gus again. He had moved to a condo off Steveston Highway, still a few minutes' drive from his former home. He and his wife continued to fish salmon on their modern fibreglass gillnetter. They have retained their vigour and joy in life, but Gus has grown more and more doubtful about the future of the fishery and the fishermen. Regulations have become increasingly onerous. There were only ten hours of fishing time for chums in the Fraser River in 2008. He and Pat had been primarily fishing near Bella Coola on the central coast for the past several years because the sockeye disappeared from Rivers and Smith Inlets. "They built a road up the river in Smith Inlet and you could see the salmon eggs in the gravel that they pulled out of the river to build it," he said. As for the once mighty Rivers Inlet, where the old canneries are now fishing lodges, "That is full of seals. We used to get paid to shoot them. Now we would get fined even when they are eating the salmon out of our nets."

While admitting that the fishermen have not been without fault in the decimation of the fishery, he wonders how his son Russ will survive. Now in his forties with a family to support, Russ works winters on the ferries. "Between us we have five boats, seven licences and fifty

Gus Jacobson with the model of the *Eva* built by Al Mason.

nets," explained Gus. "With the price of fuel you can't afford to run from one area to the other even if you do have the licences stacked on your boat. Some fishermen get a big truck to trailer their boats from the east to the west coast of Vancouver Island but that costs between $400 and $500 and that is before you catch any fish. So we keep a boat up north and another in Ucluelet."

As an indication of how frenetic and fractured the fishery has been in recent years, Gus explained that in 2008, in addition to ten hours of fishing time in the Fraser, they got four days at Nitinat for chums in the southern area. In the central area, there were three chum openings of two days each; while on the west coast of Vancouver Island, half a dozen openings for springs of a half-day each, and another three one- or two-day openings for chums in Nootka Sound. Gus explains that the possibility of a young person breaking into the fishery today is virtually nil if he has to pay for his boat and licences with money earned from fishing. While Gus paid $40,000 for the *Huki Lau I* in 1979, the current replacement cost would be more like $150,000 with another $60,000 to $70,000 each for a couple of licences to go with it.

Canadian tax law favours vertically integrated companies that have profits at one level to balance with losses from another level. In this business model, the cost of a boat and licence may be economically justified. But for a young fisherman who wants to buy into the business and take on a long-term mortgage, this is simply not going to happen. The large fishing companies have preferred ownership of the more efficient and flexible purse seiners, rather than gillnetters. Some years ago, a BC fishing company offered to trade in five gillnet licences for DFO permission to operate a fish wheel trap on a northern sockeye river. Nothing came of it at that time, but it is likely that fishing companies will continue to try to secure trapping permits on the major rivers—either directly or with First Nations involvement. All these innovations are likely to work against the interests of the gillnet fleet. It is difficult for fishermen like Gus, and younger people as well, not to see these regulatory approaches as a strategy to drive them out of the industry.

Growing up on the river has given Gus and now Russ a passionate interest in the land along the river as well. They are both part of a group of eleven fishermen, farmers and tradesmen who have leased a six-acre island in the estuary. They maintain the dykes and plant grain and corn on the island in return for hunting permission. Abiding by all Canadian bag limits, they take a small fraction of the ducks for which they provide feed. At the same time, they are enhancing the habitat for the salmon fry that find their way downriver and need transition places before going to sea. In the 1990s, as strawberry fields filled with monster houses and shopping malls, a Toronto developer turned up with what he claimed was a deed to the long sliver of land known as Gilmore Island that separated Finn Slough from the main stem of the Fraser River. He issued eviction notices to Al Mason and the others who live in the little houses that the Finns had built there.

Gus and other Finns joined the later arrivals to the little community to resist the developer and preserve the sensitive estuary habitat so important to the transition of salmon fry to the ocean. The land and water of the slough is designated "Red Zone" by the Department of Fisheries and Oceans to give it the highest level of habitat protection. An old land map was found that showed the location of Gilmore Island has shifted over the years so perhaps the developer owned a piece of non-existent land well out in the river, but no one is letting their guard down. The struggle continues, and the publicity has given the Slough a wide range of fans and supporters. When visitors come to see this idyllic spot on the edge of the growing metropolis, they are invariably delighted to see the perfectly maintained *Eva* moored at the end of the float, symbolizing so many years and careers of the Fraser River and the BC gillnet fleet. In February of 2009, Al Mason's role in preserving this bit of history was recognized by the City of Richmond with a Heritage Award. Hopefully this honour will add a little emphasis and awareness to the need to grant the same recognition to the river and its fish.

Kynoc

From Father to Son

"The best thing that could happen here would be to have an owner-operator provision," said Bill Wilson as he sat at the galley table of his boat, the *Kynoc*. Bill and his crew had just finished work on the herring seine at the net loft that he rents in Steveston. The net had been pulled aboard in readiness for the 2009 herring season. "I had some halibut quota that my son Boyd was fishing for several seasons but he didn't like the fishery and we considered selling it," explained Wilson. "I was disgusted by the way that each year I was offered so much money to lease it by the halibut buyers. They would offer me $3 per pound. They would then lease the quota to fishermen and at first only guarantee the fishermen 50 cents per pound when they delivered their catch. Later this went up to around $1.50 per pound for the boat. One company managed to lease 600,000 pounds in this way."

When he sold his quota he demonstrated his commitment to supporting his fellow fishermen. "We sold to a

Dick Yamazaki in the wheelhouse of the *Fisher Lassie* during a Gulf of Georgia herring opening. COURTESY OF YAMAZAKI FAMILY

hard-working fisherman that we often tied up alongside in Port Hardy. He didn't have the cash so we gave it to him on a four-year payout." Bill has a long-term history of innovative dealing in fisheries. In 1983, two key people, brothers Dick and Soji Yamazaki, left the company to which he and his dad Vivian had been selling their roe herring. Relationships are important in the fishing industry so Bill and his dad approached the two men. Dick and Soji said, "Well, if you guys start a company we will come to work for you." But Bill and his dad replied that they were fishermen and didn't know processing, so the four of them should form an equal partnership. That was 1982 and the beginning of Coastwide Fisheries. The start-up quickly gained a reputation for quality amongst the Japanese buyers. "In those first years it was just my dad and me who fished for them and we bought some gillnet herring," explained Bill.

Over time, other fishermen heard of Coastwide's transparency and honesty in dealing with fishermen and asked to be involved. The company's practice was to pay an advance when the herring were delivered. Only after the roe had been processed and sold did Coastwide pay the fisherman the balance. The books were open and for those who were prepared to wait, the returns were most always larger than if they had just delivered to one of the larger companies at a set price. "But some guys wanted to fish for us and the first thing they would ask for was a cash advance to buy web or some other supplies. That was how the big companies kept the fishermen in debt and delivering their catch, but we didn't work like that. We also avoided the big Japanese buyers who will give 50 percent up front but then manipulate the price after the season. We developed a line of credit by putting up our houses and our boats. We used to think, 'Man is this

The *Kynoc* built by Al Renke's Shore Boats in 1986 alongside one of the Wallace seine boats built by Menchions shipyard in 1926. The pair demonstrate sixty years of seine boat evolution.

the right thing to do?' But if you have faith in your fishery you have to do it." Coastwide thrived. After Vivian Wilson and Soji Yamazaki died, the company continued with Bill Wilson and Dick Yamazaki as equal partners.

Dick, who was born in Japan, understood the culture into which they were selling their finished product. Bill, as a First Nations British Columbian, understood the waters and the fish. It was a perfect partnership. The company attracted the top fishermen and also did co-packing for other firms. In 2007, after twenty-five successful years, Bill and Dick accepted an offer for the plant and decided it was time to take life a little easier. Dick Yamazaki plays more golf while keeping his hand in by working with his son on a salmon-based pet-food product. At sixty years of age and preparing for another herring season, Bill said, "I don't have enough outside interests or hobbies to quit fishing, but I did take a thirty-one-day Antarctic cruise with my sons this year." But even in Antarctica Bill marvelled at how the ocean's krill could be seen frozen to the bottom of the ice that the former Russian icebreaker turned over. "Once a fisherman, always a fisherman," is a true adage for Bill.

Bill Wilson's roots in commercial fishing are deep and wide. Growing up in the Heiltsuk village of Bella Bella on BC's central coast, Bill was around fish and boats from his earliest memory. Just as he was coming of age, the BC herring stocks collapsed from over-fishing in the early fishery that reduced the oily little fish to oil and fishmeal. Bill and his uncle, Wally Brown, went with a number of other BC fishermen and BC seine boats to fish unexploited herring stocks on the Atlantic coast. With his knowledge of the fish and of boats, Bill progressed from deckhand to skipper of the BC-built seiner *Quesnel Lake* while his uncle fished the *Quadra Isle* and they delivered to the packer

Boyd and Bill Wilson on the bow of the *Kynoc*.

Quatsino. "There was one bay there just a few miles from the [Weston-owned] BC Packers plant at Harbour Breton, Newfoundland. The *Quatsino* packed about 135 or 140 tons and we fished there and other places for nearly ten months of the year for six winters," he recalled while agreeing that the fishery contributed to the collapse of the Atlantic cod stocks. Like many other fishermen, Bill is perplexed by the government's inability to shut a fishery down before it crashes. "I remember fishing herring outside of Gunboat Pass [on the Central Coast] when I was just a deckhand. I had to keep pulling the net tight so that the meshes would flatten because the herring were so small they would have been able to swim out through the meshes."

Bill was by his own account "just a kid" when he became a skipper on the east-coast herring fishery. So it wasn't too difficult for him to entrust his son Boyd with running a boat when Boyd was just eighteen. "I'm an eternal optimist. So for my son, with equity in his boat and licences he can make it. But it would be very hard for someone entering the fishery without this. We have two boats with two herring licences and both A and B salmon licences [allowing them to fish the whole coast]." At the same time, Bill knows that commercial fishing will never be the challenging and competitive game that it was before quotas were introduced.

In the old herring fishery, boats waited anxiously for the Department of Fisheries to declare an area "open." They were required to stay outside a line until a committee of company reps and fishermen deemed the percentage of the roe in the test fishery to be high enough. When the roe reached about 12 percent of the body weight, the boats were allowed to rush into the area. Known as an Olympic- or derby-style fishery, the opening lasted anywhere from a half hour to several hours until the DFO felt that the allowable catch as set by the biologists had been

Father and son team. Boyd Wilson's *Fisher Lassie II* follows the *Kynoc* to the fishing grounds. Boyd sold the *Fisher Lassie II* in 2009. COURTESY OF WILSON FAMILY

caught. The best and luckiest fishermen could catch up to 1,000 tons in a single set while others went home with nothing. Byron Wright told me that in such a fishery he relied on the biggest set of air horns he could mount on his boat along with the biggest steel anchor projecting threateningly from his bow to intimidate other fishermen out of his way when he was setting on fish. It was not uncommon for the resulting disputes to be settled in the courts months after the fishery was over. It was also not uncommon for such a rapid and chaotic fishery to exceed the allowable catch.

Current practice is for the allowable catch to be divided equally amongst all the licences. Individual boats are permitted to stack two licences on a boat. This does away with the competitive nature of the fishery and slows down the harvest as the boats are virtually assured of their catch. Most fishermen, including those who were successful under the old system, agree that this is a great improvement over the Olympic-style fishery. For those fisheries that target on resident populations like halibut, ground fish and even herring this can work very well. Implementing a quota system on migrating salmon has proven a greater challenge.

Bill, who caught a lot of fish and made good money under the old system, is optimistic about the new ways. "Hopefully this quota system will work. At least we know what we are going to catch. We put freezers on the boat for the pilchard fishery and we can use those for our salmon. I can see us freezing and moving them to reefer containers ashore for storage. If we had a 1,000-fish quota and got eight hundred fish in our first set, we would hold those alongside in a collapsible pen that we made for pilchard. Then we can drop the anchor while we butcher and freeze them onboard. That way we can sell them over time at a

This classy seiner, the *Pacific Horizon*, was built by A.B.D. Enterprise of North Vancouver in 1989 and fished by John Wasden in 2009.

higher price. Last summer we got less than $2 per pound but if we sold frozen sockeye from storage later in the year we could get more like $7.50 per pound."

As a member of the Heiltsuk First Nation, Bill expressed concern about the effect of the current fishery management on his friends and relatives in Bella Bella and other First Nations villages. "What the government of Canada is doing with these buyback licences is ass-backwards," he lamented, referring to the practice to buy or create quota licences in various fisheries and then turn them over to the bands to administer. "The bands have very few, if any, boats capable of fishing the licences so the piece of paper gets leased out to white fishermen. Sometimes this has happened even if there is a band member with a boat. In Bella Bella, besides my family, there are only two other seine boat owners left and they are both older than I am." Bill acknowledged a major fishing company attempting to lease or sell boats to the villages on the condition that they deliver their catches to the company. But he stated that even among the 2,000 members of his band, it would be hard to find trained and willing crew. He was also unhappy that some First Nations communities upriver were getting commercial allocations because by the time the fish get upriver, they are dark and of a lower commercial value. "Then they sell those fish to the companies for about seventy cents per pound. The companies don't mind as they are just putting the lower quality fish into the can."

Over the years, DFO has often chartered fishing boats for various biological surveys, providing opportunities for the two professions to develop understanding and appreciation of the other's skills and knowledge. Bill and Boyd have won charters to take DFO divers on herring-spawn surveys typically lasting three weeks following the March-April spawning season, as the eggs "eye out" in nineteen

On the grounds, the *Kynoc* is ready to fish. COURTESY OF WILSON FAMILY

The *Fisher Lassie II* loaded to the scuppers with herring.

days. They go into the central coast inlets from Smith in the south to Kitasoo Bay in the north. Bill recounted surveys when the spawn has been found as deep as 15 to 20 fathoms. This work has also taken them to the inlets where the Kitasoo Band has partnered with salmon farmers to allow farms into their territory. Bill, whose boat *Kynoc Inlet* is named for the area, laments the amount of sea lice that he has seen on salmon fry from that northern area has not received the same attention as the sea lice infestation of pink salmon fry in the more southerly Broughton Archipelago.

Survival in the modern BC fishing industry has a lot to do with optimism aligned with pragmatism, and everything to do with having the licences to fish mortgage-free boats. It is not a time to be building from nothing. "My son and I are still making money," Bill explained. "We are conscious of what we spend and we do a lot of our own work." Bill has talked to fishermen in Alaska when considering selling the *Kynoc* and retiring from fishing. With the owner-operator provision keeping the debt burden off fishermen, and a more productive fishery, Alaska is where there is still a demand and money for well-built aluminum boats. Many Canadian boats under the 58-foot limit are finding buyers. There is a particularly steady demand for aluminum boats like the *Kynoc* that were built in the 1980s and '90s at Al Renke's Shore Boats in Richmond. A German-trained metal worker, Renke contributed to the development of the techniques that allowed aluminum to withstand saltwater corrosion. In British Columbia, the future of the fishery will be on the shoulders of younger fishermen like Boyd Wilson. Boyd inherited the *Fisher Lassie II* from his grandfather Vivian and continues the family fishing tradition. With thousands of years of fishing history behind him, he will work to pass onto the next generation a fishery that is better managed than the one he is inheriting.

10

The Phillips Family

From Salmon to Black Cod

Doris Phillips looked at the tiny model of her dad's double-ender salmon troller *Tarzan* and pointed out an imperfection. "It has the trolling poles mounted out at the sides," she said. "But when I was little he had their bases in at the bottom of the mast. I remember because that is where I used to sit when he was fishing. I had a long leash that gave me enough room to get into the cabin but not near the Easthope engine or the side of the boat." She shares this intimate and lifelong knowledge of boats with Ray Phillips, her husband of fifty-two years. Both grew up in the watery world of Pender Harbour and together they have owned an impressive list of boats. "In 1967 I was running a water taxi as the school ferry. Vic Gooldrup had built that boat here in Pender Harbour and we added a drum to gillnet salmon," Ray recalled when I visited them in their comfortable waterfront

Doris and Ray Phillips at their Pender Harbour home.

home. "In 1967 we built the *Surf Line*, one of the first 40-foot Pelagic fibreglass gillnet boats. Vic went on to build a large fleet of seiners for fishermen like James Walkus with his 'Joye-fleet.'"

A 40-foot boat is big for a gillnetter, but it allowed the Phillips family to move around the coast. They gillnetted salmon on the *Surf Line* off San Juan and they fished halibut out of Namu and Bull Harbour. They iced the fish and brought them into Vancouver on trips of up to twelve days. At that time, before vessel quotas, between each trip there was a mandated eight-, or at times, twelve-day layup. When they delivered to the old Campbell Avenue Fish Dock, they would phone with an ETA and their estimated catch would be put up on the halibut exchange board for the buyers to bid. While long-term relations were developed between the fishermen and the buyers, it would be a mistake to say that there was unlimited trust. Doris Phillips chuckles over the attempts by buyers to cheat

A model of Doris Phillips' father's boat, *Tarzan*, lost at sea without a trace in 1974.

on the scales and the attempts by fishermen to cheat the buyers. One old fisherman knew his boat well. When he brought in a full load of 44,000 pounds he was dismayed to find a significantly smaller number after the fish were weighed. Later he checked and, so the story goes, he found that golf balls had been inserted in the scale in such a manner that it would read less than he knew from experience it should. The Campbell Avenue Fisherman's terminal survived several decades from the 1930s until 1989 when it was shut down and demolished by the City of Vancouver. The closing of the fish dock with its several independent fish buyers also meant the loss of the legendary Marine View Café that had provided brilliant views of boats like the Phillips family's *Surf Line* and later the *Ocean Twilight No. 1*. The Safarik family, immigrant-founder John and his two sons Edward and Norman, were probably the longest-lasting tenants as buyers and traders. The hole in the waterfront remained for many years as a glaring testament to the disinterest of Vancouver city in its fishing fleet.

After the Phillipses bought the larger *Autumn Winds* they sometimes packed seine fish for Jack Grove of Bella Coola Fish. "I was a woman so the college kids that worked on the boats would try to sneak a humpy by me as a sockeye but I yelled at them and left them to mumble but with nothing more to say." At other times, Doris and Ray would gillnet for salmon according to where the runs were showing and government openings allowed them to fish. As was common practice, they had their boat rigged as a combination boat for both gillnet and troll fisheries. Since salmon trollers, being less efficient, were allowed to fish seven days a week they could gillnet as many days as permitted and then troll on the weekends. Their work ethic rewarded them with good boats and a successful family life.

Norman Safarik at the famous halibut board on the old Campbell Avenue Vancouver Fisherman's Dock shortly before it was torn down.

Good times on the *Ocean Twilight*. COURTESY DORIS AND RAY PHILLIPS

The purchase of the 58-foot wooden boat *Ocean Twilight No. 1* gave them a boat big enough for serious halibut fishing. One of the largest boats built by well-known builder Morris Gronlund on the North Arm of the Fraser River, it was also used to serve as a mother ship for herring gillnetting. They fished halibut off northern Vancouver Island and had good years prior to the introduction of quotas. They received a hefty 75,000-pound quota, and the small boat they had sold to their son Wilf, *Surf Line*, got a 30,000-pound quota. These quotas were all based on a year when BC had a 12 million-pound Total Allowable Catch. By 2009, this was considerably less at 7,630,000 pounds, from which a "sport" catch allocation was deducted and individual quotas reduced proportionately. Wisely, in the original plan, a maximum limit of one percent of the TAC was set for each licensed boat.

Ray, like most halibut fishermen, has good words for the quota system as far as it goes. "It brought some order to the fishery. When we had the shotgun fishery, where you had ten days to catch as much as you could, people would put way too much gear out to fish harder and then just cut and leave the gear when the ten days were up." This manner of fishing, called Olympic-style, also forced small boats to fish regardless of the weather and glutted the market with fresh halibut two or three times per year, while there was only frozen halibut available the rest of the time. But in practice, there are some serious flaws in the current management regime. "I would be the first to support an owner-operator provision," Ray said. "I've always believed fishermen should fish. When I am through and don't want to fish anymore, I should get out of it."

A few weeks after visiting with Doris and Ray, I was in Port Hardy when the first trips from the 2009 halibut season were being delivered. A visit to one of the offloading stations confirmed what Ray had been saying. Bad weather had forced Gary Wick in with a little less than a full load on March 30. The prices were good but his costs were high. "The bureaucracy is a big problem," Wick explained as we stood on the dock outside the little office where the validator was getting ready to review his landings. Pointing at a video camera enclosed in a plastic bubble on one of the boat's stabilizer poles he explained, "Last year the total cost for record keeping for my son's and my two boats was $37,000." The video cameras are wired so that they start running as soon as the hydraulics that haul the longlines are turned on. As each fish is brought over the side it is recorded against a coloured pattern on the side of the vessel that shows the fish's length. In addition the catch and the time that it was caught is recorded in a special logbook sorted by size and species. A complex set of provisions allow for by-catch or non-targeted species to be

Gary Wick was delivering halibut at Port Hardy in March of 2009 when he also delivered this piece of heavy trawl netting that came up with his gear.

Will Durnford heading halibut in Port Hardy off-load facility from Gary Wick's delivery.

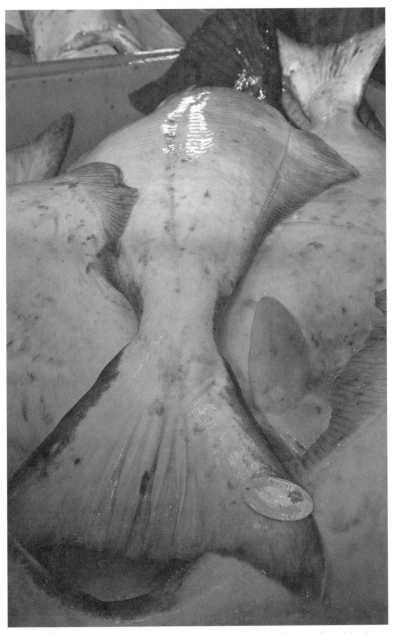

Fresh halibut just delivered from Gary Wick's *Ocean Park* in Port Hardy, March 2009. The tag on the fish tail carries a unique serial number and indicates that it has been validated.

retained. This is a popular provision with fishermen like Wick and the Phillipses, as fish brought up from depth will invariably die when thrown back—as was once done when all non-targeted species had to be returned to the sea.

There is a lot to like about individual quotas. They can allow fishermen to fish to weather and market conditions. Wick has been on the fishing boat since he was ten years old and his son was on the boat since he was five. Over the years Wick has been able to buy up an impressive amount of quota on his boat but with the other fisheries more limited his son won't be able to afford similar purchases. Another issue that rankles halibut fishermen is the logbook in which they must enter all the data of the fish caught, including species names for up to twenty different varieties of rockfish. "I can identify some but not all," explained Ray Phillips. "But the video cameras, expensive as they are, are better than having a stranger onboard a family boat." Many fishermen who cannot afford to buy quota and own only a small amount are forced to lease additional quota from "armchair fishermen" or shore-side investors. In 2008, Wick explained that fishermen were paid between $4.30 and $4.70 but many had leased quota for $3.70 per pound. The result of this was that the people who went out and did the backbreaking work of hauling the fish were earning only $1 per pound to be split amongst the three or four men and the boat. The quota owner, meanwhile, was making $3.70 per pound from what is ostensibly a public or common property resource.

All of Doris and Ray Phillips' three children were taken out on the boats early. "The rule was," said Doris, "when you have learned to swim and can cook a full meal, then you can go with Dad." Wilf and Paul continue to fish as does son-in-law D'Arcy Blake. Early on, the sons wanted to try the black cod fishery so they bought one of the forty-eight "K-licences." These licences had been defined in 1981 after some fishermen had worked to develop the Japanese market for the fish, also known as sablefish, which led to increased pressure on the stock. n 1981 the season lasted 245 days but, with the greater efficiency of the forty-eight licences, this was reduced to only fourteen days in 1989, in spite of an increase in the total allowable catch. With the support of the industry association for black cod fishermen, known as the Canadian Sablefish Association, the government introduced an individual vessel quota (IVQ) system. For Wilf and the Phillips family, this quota fishery works well. It is predictable and seems to be well managed. The only cloud on the horizon is a government-sponsored attempt to farm black cod in open ocean pens. Since young black cod rear in the very

Jim Moorehead of Longliner Seafoods on Granville Island still owned the *Ocean Wonder* back in the early 1980s. After he sold her the boat was seized for drug running. Later the Phillips family owned her and in 2009 the pretty Frostad-built boat was registered to Walter Piatocka.

Sean, Raymond and Tiffany, Doris and Ray's grandchildren, holding an 18-pound Atlantic salmon that escaped from a fish farm but was feeding in the wild. *COURTESY OF PHILLIPS FAMILY*

inlets where the farms are destined, there is a real fear that diseases may spread from the farmed fish to the wild fish as appears to be the case with farmed and wild salmon. Today, Paul fishes the big steel *Ocean Aggressor* on both black cod and halibut. With their significant quantities of quota, the family can do well in the modern fishery.

Doris and Ray are pleased with the life that they have had in fishing in spite of some trying times. In 1974, Doris's dad took his little troller *Tarzan* over to Sidney to have new gurdies installed. He called to say that he was leaving Sidney for Victoria but he failed to arrive. In spite of an extensive search, no wreckage was ever found. A pair of RCMP officers who were sport fishing off Spanish Banks reported seeing the boat inbound for False Creek. The family was left to wonder if drug runners had hijacked the boat. As the opportunities to make money and feed a family in forestry and fishing declined through the 1980s, there was an increase in drug running with fishing boats.

In 1986, when the good Frostad-built halibut boat the *Ocean Wonder* was seized with drugs onboard, I was walking on the Campbell Avenue fish dock with a former owner. A well-known fish retailer who had left his career as a fisherman behind, he declared to each person that he met, "No, it wasn't me! I sold that boat some time ago." After the drug bust the boat was confiscated and put up for auction along with its black cod licence. The Phillips family bought the boat for the licence. "It was such a mess," recalled Doris. "When I was cleaning and scrubbing I reached in behind a wall in the bathroom and brought out a bag of something that looked awfully suspicious." She promptly got rid of that evidence of the boat's former life and eventually, during a downturn in the black cod market, the family sold the boat and black cod licence to meet financial commitments. The story of

When it was declared that salmon fishing boats, such as the Phillips family boat, could no longer alternate between gillnetting and trolling but would have to choose one or the other, the government bought the resulting surplus gear. This pile of nets and trolling gurdies was a result.

the drug boat was added to an ever-growing family tapestry of stories stretching over several generations.

Ray has a long history of involvement with the fishermen's union and has put in his time on the salmon enhancement council. In the past, especially in smaller coastal communities, the role of fishermen in maintaining salmon streams has been significant. As fishermen are increasingly separated from a sense of control in the fishery, this commitment can be expected to decline. Similarly, the oversight of management decisions that many fishermen took very seriously is already in decline as fishermen increasingly become corporate employees rather that joint-venturing entrepreneurs. However, the Phillips family sees that while change is constant, there can be a good life in fishing for their children and their grandchildren, who are already taking their places on the family boats.

11
Mark Lundsten
The Political Halibut Fisherman

M ark Lundsten and his wife Teru live in a spectacular architect-designed house on a Fidalgo Island mountaintop. The roof is held up with beautiful heavy wooden timbers. The decking that runs across the front of the house providing sweeping views of Puget Sound and the Olympic Mountains is built with equally fine wood. The house and the view are just rewards for over a quarter century that Mark spent as a commercial fisherman. For much of that time, he was the owner-operator of the Seattle-based halibut schooner *Masonic*. The quality of the wood in Mark's house is reminiscent of the quality of wood that the *Masonic*'s builders, back in 1930, put into the classic 70-foot schooner. Built in Tacoma, the boat had already supported generations of fishermen when Mark bought her in 1984 and, in the seasons since he sold her in 2002, she has continued to provide a livelihood for an Alaskan fishing family.

Mark's journey, from long-haired Colorado-raised

A younger Mark Lundsten with the halibut schooner *Masonic* at Homer, Alaska, in 1986. *COURTESY OF MARK LUNDSTEN*

University of Washington grad student in the 1970s to owner of the house on the mountaintop, followed a dramatic evolution in the management of the Alaskan halibut and black cod fisheries. From part-time work in a shipyard to full-time welding to decking on a fishing boat, Mark moved through the late 1970s. As so often is the case, the boy from inland not only took to life as a fisherman but he excelled. "I had more in common with the old Norwegians who had come from another land to fish in Alaska than I had with their more conservative sons and grandsons," Mark explained when I visited him in the fall of 2009. Mark learned from the old-timers and then explored new ideas. While a lot of fishermen focus on technology to increase their catches, Mark fished traditional gear and worked with ideas to improve management of the fishery. The Pacific coast halibut fishery from California through British Columbia to Alaska and the Aleutian Islands has, since 1923, been managed by the

Mark and Teru Lundsten at their Anacortes home high on a hill overlooking Puget Sound and the Olympic Mountains.

International Pacific Halibut Commission. Canada and the US each appoint three commissioners to oversee the IPHC's work in biological research and stock assessment. The IPHC also recommends appropriate catch quotas for the various areas into which the coast is divided.

Often held up as a model of good fisheries management, the IPHC also represents an example of successful international cooperation. However, by the 1980s increases in the fishing fleet's size and efficiency were threatening the sustainability of the stocks and wreaking havoc on the markets. "Openings were typically only one or two days long," recalled Mark. "Fishermen set more gear than was needed in order not to waste even an hour of fishing time." By the late 1980s, fisheries managers in much of the western world were suggesting that the total allowable catches set by a range of scientific sampling and modelling could be divided up to individual quotas (IQ). The neoliberal economists among the fisheries managers

The schooner *Masonic* at the Pelican Seafoods plant in Pelican on Chichagof Island, Alaska. COURTESY OF MARK LUNDSTEN

often touted the additional advantage of consolidation if the individual quotas were made transferable (ITQ). But Mark and a few like-minded halibut fishermen saw that this would take the fishery out of the hands of the fishermen and deliver it to shore-based processors and investors. They proposed individual fisherman's quotas (IFQ). These later could be bought and sold but only among legitimate fishermen who were prepared to get out on the water and fish them.

It would be an understatement to say negotiations were long and protracted. Some fishermen feared the result would be an imposition of government control on their operations. Environmentalists feared that it would open the fishery to abuse. Processors worried that they would lose the control and prices that they had under the Olympic-style fishery. Mark and his colleagues countered the claims of the more conservative fishermen and the environmentalists in the initial round of discussions. Mark acknowledges that while the arguments of the first two groups were without basis, the result has been that processors lost much of their advantage. With capital invested in capacity to freeze the large volumes of halibut delivered in the short openings, they had been able to pay low prices and sell low-quality, frozen halibut to a satisfied, low-cost consumer year round. Good for them but not good for the fisherman. "Under the old regime we were price-takers, with the IFQs we became price makers," he says.

In the US system, there are management councils that present recommendations to the National Marine Fisheries Service (NMFS), the federal government agency responsible for managing a fishery. "I wrote the outline that was finally passed by the North Pacific Fisheries Council in 1989," Mark told me. "It was comprised of

Gutting black cod on the *Masonic*, with Mark Lundsten in the wheelhouse. *COURTESY OF MARK LUNDSTEN*

various points of view including my own." Mark's words serve as a short summary of nearly a decade of meetings, lobbying and debate—not to mention broken and built friendships. One point that was never really in contention was that the fishermen should control the quotas. In 1990, the Secretary of Commerce signed the agreement on behalf of National Marine Fisheries Service. At this point the processors accepted the move from an open fishery to the IFQ, but many in the Alaskan fishery fought on with the support of NGOs (non-governmental organisations) that claimed it would hurt the resource and small fishermen. They challenged the agreement in the courts, where it was upheld at the Court of Appeals, and finally the Supreme Court refused to hear the case. It was a done deal and a financial windfall for fishermen who were not only assured their share of the catch, but they also provided a retirement nest egg that they could sell to other fishermen when their time came to leave the fishery.

The amount of each IFQ was based on the fisherman's best five of the seven years between 1984 and 1990 for halibut, with a similar arrangement for black cod quotas. Each year the IFQ would be calculated as a percentage share of the total allowable catch as determined by the International Pacific Halibut Commission. Special provisions were made for "First Generation Quota Holders" that would allow them to hire skippers on their boats for some trips. The privilege would be phased out for the second generation fishermen who bought their quota shares and would be able to employ skippers in the event of medical emergency only. In the early version of the regulations, the owner-operator was defined as owning at least one percent of the vessel. "Initially there was rampant abuse of this provision," said Mark. "This led to an uproar

Crewman David McArthur gaffing black cod on the *Masonic*. COURTESY OF
MARK LUNDSTEN

Crewman Callahan McVey setting black cod longline gear from the deck of the *Masonic*. COURTESY OF MARK LUNDSTEN

and a modification to require at least 20 percent vessel ownership."

The key difference between Canadian quota management and the US model is that the intent of the US version tends to be to protect the fisherman, while the Canadian version favours processors who can purchase and lease quota. Both systems have privatized what was once a public resource, but the rights given to owners differ significantly. As a result, Canadian quota routinely sells at a significantly higher ratio to the landed price of fish as the pool of buyers is great. This makes it much harder for a young fisherman to buy into the Canadian industry. The Alaskan system also allows for quota to be broken into small units that a deckhand can purchase and bring onto the vessel with him. Four of Mark's five crewmembers had, with his backing, purchased smaller amounts of quota that the *Masonic* fished, giving them an increased share in the profits from the fishery. "They bought between three and ten thousand pounds each with my financial support. It worked very well for all of us." When Mark sold out of the industry in 2002, he had hoped that this nucleus of quota would help one or more of his crew to purchase the boat and more quota. Several attempts to structure a deal fell through and Mark finally sold the boat to one family and the quota in a dozen different lots to other owner-operators.

While Mark is justifiably proud of his contribution to the development of the Alaskan halibut and black cod IFQs, he is equally proud of one of the positive results of that system. "Once the race-to-catch era of the Olympic fishery was removed," Mark said, "we had time to respond to and solve a potentially disastrous environmental issue in our fishery." A major concern of the US Fish and Wildlife Service and numerous NGOs was the inadvertent killing

While fighting tangled gear on a rolling deck, 72-year-old Jim Clarke and Mark share a joke. *COURTESY OF MARK LUNDSTEN*

of seabirds that dove into the boat's wake to retrieve the baited hooks as the skates were being set. The focus of this concern was the short-tailed albatross. Weighing up to 20 pounds and with wingspans that can exceed seven feet, these birds had been hunted for their feathers on the few Japanese islands where they nest. By 1930, they were virtually extinct. In spite of their protected status, by the 1990s they were still on the endangered list. Should a single bird be caught on a longline hook and drowned, it could cause the closure of the fishery. "Even though it might be only a rare occurrence," recalled Mark, "we had to figure out a way not to catch any."

Mark explained that it was the IFQ system that allowed him and his crew to take part in a cooperative research project to demonstrate the effectiveness of bird by-catch avoidance solutions. They had already started to tow Tori lines behind the boat and above the line being set. The

Tori lines had a drag on the end and ribbons fluttering from their length to scare away the birds. In the open catch-all-you-can fishery with twenty-four- or forty-eight-hour openings, there was no way that crews could take the time to experiment with gear innovations to protect seabirds. In an article published in an NMFS publication, Mark pointed out an additional value to this type of work: "Cooperative research increases the knowledge of fishing communities of the processes that govern their fortunes. Well-informed fishing communities tend to be supportive of management and conservation, while fishing communities that are poorly informed tend to be reactionary and resistant."

Mark had made a video showing the fishery and the effective nature of the Tori lines that he distributed to environmental organizations to great acclaim. Not all are as politically active as Mark, but the vast majority of fishermen in both the US and Canada are intensely concerned

The *Masonic* northbound in Queen Charlotte Sound in 1996. *E. SORENSON*
PHOTO COURTESY OF MARK LUNDSTEN

with the sustainability of the resource and their way of life. Mark's film opens with an extended shot out the wheelhouse window as the *Masonic* bucks into heavy seas. Water rolls back over the deck to crash against the windows. "That was about as much as we could take—any more and we would have had to slow down," Mark recalled.

"I turned fifty hauling gear on deck," Mark reflected. "After twenty-five years, I felt that if I was going to do anything else with my life I had better do it now." After selling his boat and quota he purchased and rebuilt the *Gravina*, a former forest ranger boat built in Alaska in 1930—coincidently the same year as the schooner *Masonic*. In the summer of 2009 he and his wife Teru took a cruise from Anacortes through British Columbia and around southeast Alaska with a series of clients, but never more than four at a time. After years of hard-driving commercial fishing, Mark cherishes the opportunity to stop and explore the

The Seattle halibut schooner fleet, one of the best regarded and reported group of fishing boats in the world.

countless nooks and crannies along the BC and Alaskan coast. But he is also excited about his filmmaking and he keeps one eye on politics. "The way to do politics is the way the great Cleveland Browns running back Jim Brown did," he told me as I looked out of his study window at the expanse of Puget Sound. "He just hung onto the ball and ran straight up the middle." The Alaskan fishing community is fortunate to have had such a champion and very likely has not seen the last of him.

John Lenic

The Pilchard Cycle

At the peak of the 1936–1937 season, a record catch of well over 700,000 tons of sardines had been delivered to California ports. By 1945 when John Steinbeck's novel *Cannery Row* was published, the sardine fishery in Monterey, California, was already past its peak with 235,000 tons landed. But by the late 1940s, the oily little fish had all but disappeared from the west coast of Vancouver Island where as many as two dozen reduction plants had been rendering oil from the fish. In BC, the sleek silver fish with dark blue backs were called "pilchard," but they also go by their Latin name, *Sardinops sagax*, in both places. Pioneer purse-seine fisherman Charlie Clarke entered the industry in the years after the First World War and quickly made a name for himself during the months from July to October when pilchard were found in BC waters.

By 1929, Clarke and the rest of the pilchard fleet in BC caught 81,250 tons for reduction, 4,900 tons for canning and 150 tons for bait. It was already known that the

The *Ocean Marauder* beginning a set with the power skiff backing while the seine boat prepares to encircle a school of sardines. PHOTO BY JOHN LENIC

The *Ocean Marauder* has made a large set. The *Delta Harvester* has tied onto the corkline in order to load some of the sardines using its fish pump. Skipper Mitch Ponak is on deck while crewman Mike Relja stands on the pump. PHOTO BY JOHN LENIC

base and spawning grounds for the fish was in California and only the larger fish made it north to BC. When the catches began to disappear, the fishermen followed them south. In the days when national control reached only three miles out to sea, Canadians could work down the Pacific coast as far as Oregon. Sardines are as flighty as a flock of sparrows. In 1936, Clarke had built the 63-foot seine boat *Western Monarch* with a 20-foot beam carried well forward. But in 1941, he built a new boat to chase the fish farther offshore. "That's the 73-footer that I built like a yacht, with the narrow bow coming out to an 18-foot beam," Charlie recalled in 1983. He also put in a new two-cycle engine with a quiet clutch so as not to scare the fish, but the engine blew burning embers onto his net so he sold the boat and built the *Western Girl* in 1942. This boat worked well for him until the pilchards withdrew from the BC coast. Their total biomass continued to shrink back toward the coast of California where catches

The *Ocean Cloud* (built by John Manly in 1976) with a set of sardines at Boswell Inlet in Smith's Sound. PHOTO BY JOHN LENIC

plummeted until 1967 when a fishing moratorium for the species was implemented in California.

Twenty years later, the population began to show signs of recovery off the coast of California. Credit was given, justly, to the moratorium but another remarkable piece of research showed that there was a greater complexity to the populations. Neither the decline nor the recovery of the sardine stocks could be solely attributed to human factors. Biologists Baumgartner, Soutar and Ferreira-Bartrina explained in a 1992 paper that their studies of drill-core samples from California's Santa Barbara Basin had revealed some remarkable results. The decline of the sardine stock in the 1950s was by no means a unique event. Examining the intermittent layers of scales laid down in the sediments over hundreds of years showed that the growth and decline of sardine populations was a natural and repeating occurrence.

While little in nature is mathematically precise, the

Ocean Cloud with a set and the *Ocean Marauder* brailing on the cork line. This is off Takush Harbour, Smith Sound, with about 60 tons in the set. PHOTO BY JOHN LENIC

recurring growth and decline of sardine populations over the previous 1,700 years was clearly visible in the layers of increased or reduced scales. Working with historical data on water current temperatures and catch records, the scientists surmised that climate has affected sardines, anchovies and even salmon for at least the last 2,200 years. Further, working with fossil otoliths, the inner-ear bones, they were able to confirm that anchovies have inhabited the California coast for over a million years—though they were not able to confirm sardines beyond 7,000 years (Bowen and Grant, 1997). Far from being huge, placid lakes, the world's oceans are great swirling gyres of currents. These currents shift over time and so carry warmer or colder water closer to or farther from continental shores. Combined with deep upwelling of colder water, this can result in large shifts in the population of the ocean's plankton. It is the plankton on which the sardines feed and the sardines on which larger species like tuna can feed.

What is known is that beginning in 1990, sardines started to show up again off the BC coast. A few BC fishermen began the fishery, including the late Byron Wright with the *Prosperity*, John Lenic with the *Ocean Marauder*, Richard Leo with the *Haida Joye*, Bill Wilson with the *Kynoc*, Glen Budden with the *Ocean Venture*, Russell Arnet with the wooden *Phyllis Cormack* and Cliff Tarnoski who used a variety of boats. Their boats represented a full range of BC seine boat types and builders. Wright and Wilson had aluminum boats built by Al Renke's Shore Boats, while Leo had an aluminum boat built by Sam Matsumoto. Frostad built Budden's steel boat. Arnet's wooden boat, of Greenpeace fame, was built in Washington State in 1942 as a pilchard seiner for the earlier fishery. These fishermen applied for and were granted temporary licences by the

Dr. Don Pepper and a 10-kg box of frozen-at-sea sardines.

Canadian Department of Fisheries and Oceans (DFO) to experiment with a "new and emerging fishery."

The reduction plants and canneries were long gone from the west coast of Vancouver Island and their markets for fish oil and meal had also gone when the sardines returned. It is one thing to have a fishery and quite another to have a viable market for the fish. The fishermen who pioneered the renewed sardine fishery (the term pilchard was no longer in use) were determined not to allow the fish to go to a low value reduction process. In the years since Charlie Clarke and others chased the declining pilchard schools south to the Oregon coast, BC fishermen had become much more sophisticated. It was not unheard of for a fisherman to fly to Japan to research markets for other species. Now the highest-value market was found to be frozen-at-sea (FAS) 10-kilo boxes of perfect fish in the large 200-gram size found only in BC. The California fish tended to be in the 80-gram range. The sardines' blood can turn bad soon after they die, causing the fish to spoil quickly. A sardine that has not been cooled quickly or frozen will have a reddish eye and gill cover. The Japanese buyers insisted that the sardines supplied by BC fishermen must have no red-eye and barely a scale out of place. They had to be hand laid in the boxes in perfect order. It was tiresome work for BC fishermen, who traditionally have left such tasks to shore workers.

However, a price per ton in the $2,500 range helped ease the pain. The BC product quickly gained a quality reputation, and not just in high-end Japanese restaurants. The product was also destined to go right back onto fishing boats in Japan as bait for big eye tuna. The fishermen, with fisheries economist Dr. Don Pepper, formed the Canadian Pacific Sardine Association to work with the DFO in planning the fishing season (normally from

September to November), as well as to market their catches. BC sardines now, in addition to being tuna bait, go to the Philippines, Russia and Ukraine as food and also to tuna farms as feed. The price has come down somewhat but there is still money to be made under the right conditions.

The development of the sardine fishery and the association represent some of the best successes to the modern BC fishing industry. When Pepper and Byron Wright were teenagers in Alert Bay in the 1950s, naturally they both went out on the seine boats with First Nations skippers. They fused what they learned from these skippers with the European ways they got from their parents and the universities that they both attended. In time, Byron became one of BC's highest-earning salmon and herring seine skippers while Don attended the London School of Economics before earning his Ph.D. in economics from the University of Wales. He worked for DFO before beginning his twenty-year career teaching economics at the BC Institute of Technology, from which he is now retired. But the fishing fever never left him and he regularly went out seining with Byron, using his statistical skills to predict when and where the salmon or herring were most likely to show. An early adopter of computerized sea surface temperature studies, he annually helped Byron decide if the Fraser sockeye would return down the inside of Vancouver Island or around the southern tip.

Also in the original group that pioneered the fishery on the revived sardine fishery was John Lenic. Unlike Byron, John grew up in a fishing family. His dad had immigrated to Canada from Croatia in 1929 and crewed on the early pilchard boats as well as on the Gulf of Georgia Japanese-Canadian herring pair seiners in the 1930s. Much of the Croatian fishing community was concentrated in the

John Lenic with the *Ocean Marauder* (built by John Manley in 1978) and powered by a Cummins KTA38 main engine.

Burnaby Heights area above Second Narrows. It was and is a community in the truest sense of the word. John recalled, "My dad told me 'Be anything that you want, but don't be a fisherman,' but here I am a fisherman. So when one of my sons was fourteen years old, I took him 1,400 miles offshore tuna fishing." Neither of John's two sons or his daughter are fishermen.

The Croatian fishermen were strong contenders among the coast's purse-seine fishermen. Within the community John had plenty of successful role models whom he continues to learn from after decades of experience. "Marcelo Carr told me about being up in the crow's nest on those pilchard boats," said Lenic recently of a conversation the two had about the days long before the advent of sonar. "He said to look for a dark spot on the ocean surface. If there is no cloud above to cast a shadow then that is pilchard." Carr also told Lenic of looking into the ocean swell to see "breezers" or schools of sardines breaking water.

Lenic's *Ocean Marauder* doesn't have a crow's nest but he makes up for that with an amazing array of wheelhouse electronics from radar, several phones, global positioning, sophisticated colour depth sounders and—most importantly for sardines—a sonar. "I set my sonar at ten degrees downward tilt to find herring, but sardines are right under the surface so for them I set it at only a two-degree tilt," he explained.

Just as Charlie Clarke spoke of building a boat that could "sneak up on the fish," Lenic has learned of the speed that sardines can travel. "Some fishermen think that sardines dive under the purse seine's lead line," Lenic explained, "but they don't. It is just that when you set ahead of them the lead line doesn't sink fast enough to block them. Once I was on a school of sardines that were swimming away right on the surface. My GPS was showing that my boat was making 5.3 knots and just keeping up with the fish, so they are fast swimmers. As Lenic speaks of the trials and tribulations as well as the joys of catching sardines, a tone of excitement emerges. It is thrill to be heard in the voices of all humans who pit their will and their skill against fish. It is as familiar to the dry fly fisherman on a mountain stream as it is to the skipper of the largest ocean-going tuna seiner.

His voice takes on a significantly different tone when he speaks of the management of the fishery that he and six other fishermen developed. He explains how in time they were required to accept the expansion of the fleet. There was also an agreement to restrict nets to the same 225-fathom length of a herring purse seine. This was not an onerous restriction as most fishermen were already using their herring seines. Nor did Lenic mind the expansion but, "Why fifty licences?" he fumed. "What is that number based upon? Last year only seventeen boats fished and most of them had to lease a second licence to make it economically viable."

The need to "stack" or lease two or more additional licences was prompted by another number Lenic feels to be arbitrary. To calculate an optimum catch for a fishery, the managers estimate the total mass of fish. From that, they calculate a sustainable harvest. In the 2008 season this was set at 12,000 tons. "But only seventeen boats fished," Lenic explained, "so even with up to five licences stacked on a boat there were only 10,000 tons caught." Lenic estimates the mass of sardines to be in the neighbourhood of a million tons. "I turned my sonar on off Estevan Light and left it on for about 80 miles up to the buoy outside of Ucluelet. My sonar reaches out for a half-mile radius so the full circle of its reach is about 600 acres. There are a lot of fish out there."

The *Delta Harvester* on a set outside Rivers Inlet in 2009. The skipper was Mitch Ponak and owner was Wayne Martinolich. The boat has recently been sold to Randy Reifel. PHOTO BY JOHN LENIC

But reports like Lenic's are often called, somewhat dismissively, "anecdotal" by government managers. This is a feature not only of Canadian management but of fisheries management throughout much of the world. Part of the problem is that in cultures trained to rely on firm numbers, it is often only the highly regulated landings that are quantified. Observations by the most knowledgeable fishermen whose livelihoods are dependent on accuracy tend not to be given the weight that they are due. At the same time, those who have given the most thought to the fishery know that predicting amounts of fish in a large area of ocean is somewhat of a guessing game. Lenic maintains that a superior system of management would permit boats to catch sardines in 50-ton increments with another increment being granted after each delivery until the quota had been taken. Don Pepper stated that "Sardines are like a gold mine—they are where you find them."

"Catching the sardines is the easy part," Lenic explained. "Once you get them on board, you have to care for them." In the early fishery he, like others, processed the packaged frozen bait on board. He despairs over the lack of interest in developing a proper infrastructure to support shore-based value-added processing in the BC industry. Today, much of the sardine catch is purchased by US buyers and taken south for further processing and then exported. More recently, he has been chilling the fish and delivering them to Port Hardy or Ucluelet. It is important to cool the fish from the fourteen- or sixteen-degree waters that they favour to between one degree and zero Celsius as quickly as possible, due to the blood condition explained earlier. Most modern boats no longer use ice for chilling fish, with refrigerated salt water being the most common. The fish are moved from the net to the hold with a large mechanically operated scoop net or brailer, or

A large bag of sardines in the seine at Boswell Inlet. Spencer Serka is the skiff man. *PHOTO BY JOHN LENIC*

they are pumped with a pump developed in BC by Matt Skikich. The pump is lowered into the mass of fish and sends them up through a large aluminum pipe. An elbow in the pipe is formed with longitudinal aluminum rods to allow the water to escape overboard while the fish go into the fish hold. In a constant quest for improved quality, Lenic planned to modify his de-waterer that had been designed for 100-gram herring so that it would help maintain the quality of the typical 200-gram BC sardine. An adage that all good fishermen adhere to is that "You can't add quality to fish— you can only preserve it."

The return of the sardines has also heralded the return of humpback and other whales that are now frequently sighted inside Vancouver Island. Offshore, the sighting of a pod of feeding whales is often used as a sign of fish. Another interesting phenomenon that fishermen have observed is the mass die-off of "lost" pilchards. As summer progresses, the sardines work their way up into Smith, Seymour and other deep coastal inlets. As the fall and early winter reduce the deeper water temperatures, the fish in the inlets will stay in the warmer water and apparently become confused to such an extent that they eat all the food in the confined area but can't find their way back south. Eventually they sicken with a virus and die, collecting in a stinking mass on the surface. This process remains something of a mystery, along with the predictability of their population cycles. But in the galleys and wheelhouses of the BC seine fleet, there is no end of speculation about their lives and the length of the current upswing in their population.

REFERENCES

Baumgartner, T.R., Soutar, A., and Ferreira-Bartrina, V. (1992). Reconstruction of the history of Pacific sardine and northern anchovy populations over the past two millennia from sediments of the Santa Barbara Basin, California. California Cooperative Oceanic Fisheries Investigative Reports #33: 24–40.

Bowen, B. W. and W. S. Grant (1997). "Phylogeography of the sardines (*Sardinops* spp.): Assessing biogeographic models and population histories in temperate upwelling zones." *Evolution* 51: 1601–1610.

13

John Radil

On Hake Trawling and Ocean Ranching

In their first generations in Canada, many immigrants worked a fishery that replicated a fishery in their home waters. In this fashion, Norwegians took up halibut fishing in the rugged north Pacific reminiscent of the Atlantic cod fishery off the northwest coast of Norway. Similarly, the Croatian immigrants took to seining salmon and herring in British Columbia's coastal waters similar to those of their ancestral Adriatic Sea. Albert Radil's father, who arrived on the BC coast from Croatia in the 1920s, went to work on a seiner and, in time, built the seiner *Good Partner* at Fenner and Hood in 1936. Albert followed his father into the industry and built the *Theresa 1* at Harbour Boatyards in 1944. Both of these wooden boats are still active on the Vancouver register. There were other boats in the family but it was the building of the *Canadian No. 1*, 26.8 metres in length, at Bensons in 1964 followed by the

28.59-metre *Royal Canadian* built at Mercer's Star Shipyard in 1968 that began the family move to other fisheries.

In 2009, sixty-one-year-old John Radil, the eldest of three brothers who make up the third generation in Canada of the Radil family, recalled that he had his first season earning a crew-share as a fifteen-year-old in 1962. When the reduction herring fishery was closed in the late 1960s, the family decided that they didn't want to follow the rush to fish herring on the east coast as many were doing. Instead they would take the big new steel boats into the trawl fishery. While there was a well-established trawl fishery in BC, it tended to be marginalized by the halibut longline and the salmon or herring seine fisheries.

The plating on the *Royal Canadian* had been built a little heavier to handle the heavier gear in the dragging fishery, but both boats proved their worth. Eventually a

The Radils' boat, the *Royal Canadian*, running down Johnstone Strait with a load of hake. PHOTO BY JOHN LENIC

The Radils' *Canadian No. 1* alongside at Steveston, BC.

third steel boat, the 23.8-metre *Sea Crest* built in 1979, was added to the fleet in the 1980s. "We still have herring licences," John Radil told me in March of 2009. "In fact one of our boats just came in from the herring a couple of days ago." But John and his two younger brothers, George and Albert, are known on the coast as highly successful draggers. Their decision to stay and develop that fishery while others went to the east coast has worked well for them. The Radils like to fish Queen Charlotte Sound rather than the west coast of Vancouver Island. "The hake seems to have less of the enzyme [that can cause the flesh to go soft]. It works well as it usually takes us about twenty-four hours to get the fish and then another twenty-four hours to deliver it to Vancouver. Putting the fish into slush ice with coolers we can actually keep it for three days." Today, explained John, "Ninety percent of our catch is hake that we get with our mid-water trawls. We work the three boats together as a pool so that we can maximize our use of

John Radil at his home in Richmond, March 2009.

the vessels." The Pacific hake fishery was certified by the Marine Stewardship Council in October 2009.

They may have sold the salmon licences, but John still credits the salmon fishery with his family's start in Canada and in fishing. In 1983 as I began to collect data on BC fishing history, I bought the Radil family collection of old *Western Fisheries* magazines from John's mother at the family's Trinity Street house in Vancouver. They were stowed in old wooden pop boxes and were a useful find, but the young Radil brothers, for various school projects on commercial fishing, had cut out most of the best boat pictures. This is a family born and educated in the role of salmon in British Columbia economics and culture.

"A big part of my resentment about fisheries management," John told me twenty-five years after I bought the family magazine collection, "is about the salmon. There are half a dozen issues that they could be addressing. It is remarkable how these issues keep coming back with little or nothing done to address them. Last year I was in Port Simpson and I talked with the First Nations there. They wanted to try to do ocean ranching in a stream that they have but DFO wouldn't allow it. Well, it works in Alaska and it is wrong not to let them do it."

John's comments reminded me of a trip that I made to Bute Inlet in the early spring of 2000 to look at an ocean ranching project that the Homalco First Nation was trying to get going. A number of BC fishermen have, in recent years, travelled to Alaska to look at the community hatchery model that has helped take Alaska's annual salmon harvest from less than 50 million to more than 200 million over the past twenty-five years. They have been impressed by what they see. Investments like community-based hatcheries with rank upon rank of incubation trays in huge warehouse-like buildings are producing fry that will be reared in the native stream waters to imprint them so that they will return to that stream. More significantly, the Canadians see a state government committed to maintaining its commercial, subsistence and sport

John Radil's model of the *Royal Canadian*.

The Homalco Salmon Enhancement Project facility at Orford River in Bute Inlet. There is now some hope for its revitalization. *COURTESY OF THOR PETERSON*

fisheries along with the coastal communities that depend on them.

BC fishermen who visit Alaska also see area-based management that gives real authority to the communities. Alaska's Private Non-Profit Hatchery Act of 1974 subsidizes hatcheries in the initial years while allowing them to harvest and market surplus returns to support their operating costs. The hatcheries restock over-fished streams. The operators pen smolts in target commercial and sport fishing areas to operate as preservers of the wild stocks on which Alaska—which has forbidden salmon farming—is building its future. What sparked the Alaskan model was a 70 percent decline in Alaska's salmon harvest between the 1930s and 1970s. The Alaskan government began restoration work and the fishermen demanded a role. Fearing corporate domination, they argued against private aquaculture, and ocean ranching for profit as had been proposed in Oregon. They pushed for non-profit regional associations that allowed them input into policy decisions. With enabling capital to found the associations coming from government, five regional associations were formed in the first stage of the recovery of Alaskan salmon. Next, fishermen sat down with their government counterparts from the Department of Fish and Game to develop regional plans that included harvest targets equivalent to the 1930s statewide level of 100 million fish. These targets have now been met and doubled. Commercial fishermen pay a small tax on their landings to help support the hatcheries.

Until the 1980s, a good proportion of the chum salmon taken in the Johnstone Strait seine fishery were returning to the mainland inlets, including Bute. In coho season a big fleet of commercial trollers was based in the mouth of the inlet at Stuart Island and local gillnetters worked well up in the Inlet. But this is no longer the case. For nearly a quarter century there has been no commercial fishery on the inlet's stocks. At the mouth of Bute Inlet, around the Yuculta Rapids is a cluster of corporate sport fishing lodges belonging to outfits like Boeing Aircraft, Paccar (makers of Kenworth trucks), Dennis Washington (Montana-based owner of railways and towboats) and Ritchie Brothers (international auctioneers of heavy equipment). Even these are suffering the shortage of fish.

Possibly the biggest loser has been the Homalco First Nation, whose village Church House was located just south of the inlet mouth. With Canadian fleet reductions, most of the village's boats were lost and with them the few jobs. Now the village is deserted with the people living on new reserve land at Campbell River. The inlet that once supported a rich commercial, Native and sport fishery is now fighting for survival. The fisheries' collective fate hinges on the cooperation of all three sectors—and on the support of government officials. That's where things get dicey. In 1991 the Homalco First Nation had applied for and received limited five-year funding under the Canadian Department of Fisheries and Oceans Aboriginal Fisheries Strategy. The $250,000-per-year grant provided for a small chum salmon hatchery to be set up in an old logging camp on Homalco reserve land beside the Orford River estuary halfway up Bute Inlet. The land, 680 acres, was designated Homalco Indian Reserve No. 4 in 1888. Ironically, it was recognized in a government report during the designation that "Large quantities of salmon are obtained from the river"—even though future government policy did not pursue the protection of this asset.

In the 1920s, a short logging railway was built from tidewater up through the reserve to haul out the old-

The Orford River, like so many rivers on the BC coast, was logged without mercy and the salmon runs were destroyed. Now, as forest cover returns to the valley bottoms, the time has come to restore the runs.
COURTESY OF THOR PETERSON

growth timber from the flat valley bottom. In the ensuing years, millions of dollars worth of timber have been taken from the valley and more recently from the steep valley sides. In Canada there has been a move to construct artificial spawning channels parallel to existing natural streams to maintain optimum water flows and gravel sizes. A number of these have been successful, but replicating nature is difficult. In the Orford, such a channel was attempted in the 1990s. With about a mile of spawning gravel, the channel looks great, but silt brought down by the river has filled the spaces between the gravel like masonry grout. A deep settling pond at the head of the channel sits untended and full of silt. Built at great cost, the channel represents one more example of the half-hearted Canadian attempts at restoration. The demise of salmon fishing in Bute Inlet is a familiar story, replicated in other inlets. At Rivers Inlet, where Oweekeno River and Lake

A vessel transporting farmed Atlantic salmon passes a purse seiner fishing for wild sockeye salmon in Johnstone Strait.

once supported huge sockeye runs, the grizzlies once fattened on sockeye now starve.

By 1995 the Homalco Band recognized that the little Orford River chum salmon hatchery, like so many others along the coast, represented little more than a token effort to restore the heritage from which they and the salmon had grown. "We had gone through a steep learning curve in managing the hatchery," explains Thor Peterson, an advisor who worked with the Homalco, "but we realized that we needed to expand to coho and chinook if we were really going to restore the Bute Inlet fishery." The Band's advocacy hit a granite bluff when, returning to Canada from an Alaskan visit, their leadership met with DFO officials. Pointing to the "common property" nature of the resource, the Canadian government saw the Alaskan approach as "salmon ranching" and a privatization of the common property resource (apparently this was not a problem when assigning quotas in various fisheries).

Although disappointed with the Canadian government's reaction to their enthusiasm, the Homalco project supporters managed to gain permission to approach private funding sources to extend their operation to include coho. The response from a single donor was enough to get the operation off the ground. However, the years since 2001 have seen a typical scenario played out. Political changes in the First Nation combined with limp support to outright opposition from the federal and provincial governments has left the facility in a quagmire of indecision and very limited success. Meanwhile, two initiatives supported by the Province are going ahead with considerable speed. Initially, five Atlantic salmon farms were approved for the inlet. Considerable local opposition had the plan reduced to one at the mouth of the inlet where the wild fry will still have to swim past. At the same time, the province is promoting a huge power project for the Homathco River farther up the inlet. Private foreign investors support both the salmon farms and the hydro projects but this time there is no outcry from either level of government about the privatization of a public resource although now it is clearly for profit.

John Radil is a patient Canadian patriot so he believes in trying to give the government the benefit of the doubt: "The [Fishermen's] Union says that there is a conspiracy to kill the salmon fishery but I don't think that the government is smart enough. It is the same old stuff that made my dad frustrated because they wouldn't listen. It's not worse, it is just more of the same but maybe there was a little less bureaucracy before." Anyone who attended a meeting with DFO and fishermen back in the 1990s might wonder. The hapless DFO bureaucrats would rent a hall in a hotel in a fishing community and a hundred or so fishermen would listen respectfully to their presentations while the DFO public relations people hovered nervously on the sidelines. Then the question period would begin and men and women who had spent decades battling huge seas and unruly nets would try to communicate what they knew to be the reality of coastal fisheries. The trained bureaucrats would stick to their agendas and repeat their message while occasionally offering to "take your concerns back to Ottawa." Finally a fisherman, no longer able to contain his frustration, would shout, "I've been coming to these meetings for twenty years and you claim to listen but I've seen no evidence of your listening."

And then we would all go home and the fishery would be a little smaller and a little more restricted. The Radil brothers continue to attend such "consultative" meetings and the fishery continues to support them. But John can't help but join other fishermen who recall how the pit lamping of herring was finally banned in the late 1960s after repeated requests by commercial fishermen. Now when they travel at night past a salmon farm, John routinely sees bright lights that are ostensibly to aid the farm workers but are also doing a fine job of "attracting krill along with juvenile salmon and herring to feed the farmed Atlantic salmon and reduce the foreign corporate owners' feed bill. It's almost enough to make someone want to yell at a bureaucrat."

Unfortunately a lot of those fishermen who were speaking so strongly for the fish back in the 1980s and 1990s have been driven out of the fishery or, in other cases, have taken the windfall gift of licences and quota which they are allowed to lease to others. But some others, like John Radil, maintain a passionate connection to the fish that nurtured them in childhood, paid their mortgages in adulthood and sustained them and their families over the decades. They even carry enough of the fisherman's optimism to feel pride when younger family members, like John's nephew George Radil, Jr., begin their first years running a vessel and add a generation to a proud Canadian tradition.

14

Truong

The Immigrant's Story

For many people living in South Vietnam after the defeat of the US forces in 1975, there was little hope of a future for them or their children. As a result, many chose to make the extremely dangerous flight from their homes in small fishing boats. There are many stories of heroic voyages across the Gulf of Thailand to refugee camps in Thailand and Malaysia. The challenge was to escape the Vietnamese authorities and then avoid the pirates who waited to rob, rape and kill the refugees. Lon Truong would face this challenge with the help of a friend. Lon's family owned a sugar mill in the Mekong delta southwest of Ho Chi Minh City. They also owned a diesel engine repair shop. From the shop, Lon took a small two-cylinder engine and installed it in a boat he had built. He then applied for and received a permit to deliver fish sauce from a factory on Dao Phu Quoc Island 75 kilometres off the west coast of the delta. The small factory belonged to Lon's friend Tran Van Tam's father-in-law. Lon began his

Lon Truong and his son Jackie in the wheelhouse of the *Debra Mar.*

Lon Truong's boat, the *Debra Mar*.

voyages to the island in 1977 and continued for a couple of years until the patrol boats became accustomed to seeing the familiar boat on its routine trip.

In 1979, the time to escape had arrived and Lon pulled the little two-cylinder engine from the boat and replaced it with a powerful 800 horsepower V-12 engine. Loading forty-two people including his family into the boat, he made as if to do his usual run then, once clear of the patrol boats, he opened up the powerful engine that pushed the boatload of refugees to safety. Lon and his cargo of refugees passed by Dao Phu Quoc and turned northwest into the Gulf of Thailand. Passing offshore of Cambodia they continued on to the southeastern Thai province of Trat. To be safe, they also continued on past the tourist islands of Ko Kut and Ko Chang to the neighbouring province of Chanthaburi. Lon had left Vietnam in the evening so it was about 4 am when, after twenty-nine hours at sea, he drew near the Thai coast. He could see the lights of the fishermen's boats. It was amongst them the pirates were reported to be lurking so he lay offshore. With daylight he landed safely in eastern Thailand.

After eleven months in a refugee camp, Lon and his family were granted permission to come to Canada. He was placed in the Kootenays area of BC where he was unable to find work. Eventually he found his way to Toronto where he worked for six years as a welder. Meanwhile, his friend whose father-in-law had owned the fish-sauce-making plant had found his way to Vancouver. After working four years in Canada, Tran Van Tam, who had skippered a fishing boat in Vietnam, had saved enough money to buy a smaller wooden salmon gillnet boat. Like immigrants from many nations before him, Tran fished his wooden boat and made money. He fished the boat hard and when it began to show signs of wear he took his savings and

The *Miss Jane II* is sister Deltaga to Lon Truong's boat

bought a modern 39-foot Deltaga fibreglass boat. With the freedom to work provided by commercial fishing, Tran had built a small fishing business. To make a little more from his catch of shrimp, he sold most of it from the deck of his boat at the Steveston fish sales dock.

He encouraged Lon to come to BC and try his hand at fishing. Although Lon had never been a fisherman, he had worked around boats and made the move. Arriving on the west coast in 1987, he arranged a loan through the Royal Bank and bought a sister ship to his friend's Deltaga. These are good big boats and, in addition to their value as salmon gillnetters, they make a good platform for shrimp beam-trawling. Lon was, like Tran, able to get a boat with both licences. As difficult as it must have been to take that earlier boat out into Vietnamese waters back in the 1970s, Lon found the BC coast to be a whole new world. "Over here it is different," he told me. "There are many

The Steveston fish sales dock, like others in False Creek and Nanaimo, have been a boon to independent fishermen and to the consumer, as it allows independent fishermen to get a better price and the consumer to get excellent fresh fish.

Barrie Farrell-built fibreglass *Debra Mar* registered in 2009 to Hoang Tan Nguyen bucking into a westerly in the Gulf of Georgia off the Fraser River.

islands and rocks and more tides in places like Seymour Narrows."

I had first met Tran in 1989 when he was raising four children and making his way in the western world. He has since sold his fishing business to begin a small business ashore. In 2009, Lon Truong continued to fish. After fishing salmon for ten years, he sold his salmon licence and repowered his boat. He has continued to fish shrimp and, as Tran did, he sells them from the dock in Steveston. It is a hard life. Lon fishes alone trawling at depths from 85 to 110 fathoms and staying at sea and freezing his catch for up to twenty days at a stretch. "It gets boring sometimes," he told me of his solitary fishing life as we sat in the galley while his wife and children sold shrimp on a busy sunny April Sunday. "There is no one to talk to. I want to fill up and go home early."

Fishing alone halfway up the coast can also be dangerous. Lon's heavy trawl net is held open by a hefty 53-foot aluminum pole. He carries a spare and when the pole breaks on some rough bottom he has the skills and equipment to weld it on board. When his daughter Julie was young and before his son Jackie was born, his wife and baby sometimes went with him. Now that there are two children going to school, that is no longer possible. Instead, they share family time on board the boat while selling shrimp in Steveston and at their Surrey home between trips. The Steveston dock is a boon not only to fishermen who want to make a little more for their fish but also to the consumers who appreciated fresh seafood and affordable prices. Lon says that the majority of his customers are from places like Taiwan, the Philippines and China. "When I first started everyone wanted only fresh shrimp but now they understand that the frozen-at-sea shrimp is very high quality," he explained. The sales dock,

Thuy Dang and Lon Truong selling shrimp from their Deltaga fibreglass boat, the *Debra Mar.*

like others in False Creek and on Vancouver Island, is also an opportunity for families to see something of where their food comes from and to shop for healthy wild fish in their home communities. When the dock was put in a few years ago, a regulation was made that fish could only be sold by the person who caught them. In the slow season for dockside sales, Lon stores some of his product at a freezer facility in Cloverdale. He also sells excess product to Lions Gate Fisheries.

Lon takes the view of many immigrant fishermen before him that his hard life as a fisherman will pave the way for a better life for his children. "It is usually okay for immigrants because we know how to work hard even for four dollars a day. But the kids don't fish. It is only me who has to. As a parent you just help them to be good people." Lon hasn't forgotten his Mekong Delta roots and on return trips he has bought a mangostein orchard and

Jackie and his sister Julie Truong on the family boat.

built a house that he talks of retiring to one day. At fifty-two years of age he has spent over half his life in Canada. Now, when he visits Vietnam, he needs to use an air conditioner as he is no longer accustomed to the heat. Many years ago when my own immigrant father tried returning to England after a couple of years in Canada it was the thorn of a devil's club irritating his elbow that reminded him of the Vancouver Island river valleys that were now his "home." If and when Lon gets to return to live in Vietnam, perhaps a bit of skin pierced by a shrimp spine will bring him back to the coast that he now knows better than most native-born Canadians.

15

The Assu Family

Sustainable Fishing

Herb Assu was my father-in-law at the time I fished on his boat in the 1960s and '70s. In the years that I knew him, I learned that he cherished two things above all others. The first and perhaps the greatest, was to make a set on sockeye jumpers at The Slide, a place of tricky tides in lower Johnstone Strait. The way that Herb set his net with the little 57-foot *Departure Bay No. 3* and worked that 1,200 feet of net in the tide, with only the old Cat 77 thumping out its few horsepower, had all the beauty and majesty of the captain of a square-rigger setting his sails to fill with air. Herb had many other sets that he could make in the waters that his father Dan and his grandfather Chief Billy Assu had also seined, but The Slide was special. The other thing that Herb cherished was to have his family working on the boat with him. A typical crew in those days would be his wife Mitzi, his uncle Jack Quatelle, his two teenage sons Harvey and Wayne, and his son-in-law, me. He would pay me by the hour to help paint the boat

Herb Assu's boat, the *Departure Bay No. 3*, about 1982 when it was still a table seiner.

---

at the old Nelson Brothers Shipyard at Queensborough and, in the final days of cotton seines, to help build the net at the St. Mungo net loft where the south end of the Alex Fraser Bridge is today.

When I started with Herb, I was an eighteen-year-old with typical insecurities and without a high school education. That first year, after the salmon season, he took me out on the winter reduction herring fishery and I learned about wind and snow in Hecate Strait. Although I eventually completed a bachelor's degree, the education that has served me best was that which I got on the deck of Herb's seiner. When I left the boat for the last time in 1973, I took with me a lifelong debt to the salmon and to the Assu family for the opportunities to learn that they had given me. During the years that I fished, I watched Herb's

Daryl Assu working the brailer handle.

son Daryl grow from a four-year-old toddling around the deck to a strapping young half-share teenage deckhand. In the years after I left the boat, Daryl developed his dad's calm, self-assured manner. When Herb died, Daryl was ready to take over the family boat and with his younger brother Mark he has continued the long established Assu family tradition of fishing the tides at The Slide and along the BC coast.

By the time that Daryl took over the business in the 1980s, some major changes had come to the fishery. The gear revolution of the 1950s and '60s, especially the drum seine, had dramatically increased the number of sets that a boat could make in a day. Increased horsepower and nylon nets made it possible for less-skilled fishermen to set in the tricky tides of places like The Slide. The increased efficiency had led to fewer fishing days and more intense competition. The introduction of a limited-entry program in the early 1970s was intended to cap the number of boats, but various provisions such as one that allowed a

Mark and Daryl Assu watching for jumpers at The Slide, which was also their father Herb Assu's favourite spot.

number of small gillnetters to be retired and replaced with a new seine boat caused the fleet to increase in number and efficiency.

The generational knowledge Herb had passed to his son of the fish and tides at places like The Slide was no longer indispensable. Before the limited entry and gear modifications, the fishing companies had competed with each other to keep fishermen who could provide fish from a certain locale. After the changes the knowledge was no longer as important. As the price of the limited licences climbed along with the advent of more sophisticated steel, aluminum and fibreglass boats, the value shifted from the crew to the vessel. This shift was not lost on the fishing company accountants who had carefully encouraged debt commensurate with the fish deliveries for the vessel owners.

As limited entry and capital improvements increased the costs, debt was shifted from the fishing companies to the banks. In the 1980s and 1990s, the cost of a salmon

The *Splendour* was built at Ladner in 1940 by Mario Tarabochia for the Martinolich family.

licence was often influenced by the success of the previous roe herring season, which was the most likely source of purchasing funds for fishermen. But many fishermen used the licences that they had inherited from before the limited entry was introduced. As the herring fishery divided the coast into three areas for licensing, a fisherman could get a licence for a second area by contracting his first licence to the fishing companies. While the old system of paternalism and company-store style support was weakened, many fishermen still found themselves bound to the companies through contracts that advanced operation costs and provided services like net lofts.

Daryl Assu found himself pressured when a company that held his debt was sold to another company and he was obliged to sell his salmon licence to escape the new company's control. Such negotiations can be abrupt and blunt and leave little room for the fisherman to manoeuvre. Daryl took the only route that he could see and so ended up with the seiner *Adriatic Star* and a single herring

Daryl Assu now owns his father Herb Assu's last boat, the *Adriatic Star,* seen here coming into Campbell River with Quadra Island in the background in the 1980s.

Making a set at The Slide on the *Splendour*.

licence. This allowed him to get by and his knowledge of the tides allowed him to skipper other people's seiners on some salmon openings. When his First Nations band put up a community salmon licence to be used under a lottery system, Daryl's younger brother Mark won the lottery and got use of the licence for the year.

It was the fall of 2001 and I had not been fishing in Johnstone Strait for some years, so I jumped at the chance to go with Mark and Daryl on the Martinolich/ Tarabochia-built 1940 seiner *Splendour*. In many ways it was just like the old days, as we travelled north on a boat that was built before any of us were born. Experiencing the dog-salmon fishery was at once exhilarating and sobering. Change may be the fishing industry's only constant, but more had changed than most Johnstone Strait fishermen can comfortably accept. In 1973, the chum salmon openings were typically two days with an additional twenty-four hours added to give a three-day fishing week. In 2001, the first dog-salmon opening of the year was announced for Monday, October 1. It would extend for only twelve hours, from 7 am to 7 pm. Thirty-four-year-old Mark Assu, who was the skipper on this trip, left the dock at Campbell River Saturday night and passed through Seymour Narrows with a fair tide lending speed and rolling the 60-foot *Splendour* in a comforting embrace. In the early '70s, Herb Assu would have left the dock on a Sunday morning for an opening that invariably came at 6 pm on a Sunday. Now most of the fleet would spend the Sunday scouting for fish and then picking a spot and holding position a good twenty-four hours ahead of the Monday morning opening.

Once we were through the Narrows, a low fog settled over Johnstone Strait leaving a bold harvest moon showing over the boat's mast but severely limiting low-level visibility. In the wheelhouse Mark and Daryl told stories of losing the radar in such conditions and navigating by following the five-fathom contour on their sounder. I recalled travelling with their father in the fog at night before our boat had radar. But we also didn't have huge Alaska cruise ships looming out of the fog like floating towns. Two hours after passing through the Narrows, Mark and Daryl dropped anchor just off Johnstone Strait in Blind Channel, a good spot to fish dog salmon bound for the mainland inlets.

I climbed into a fo'c's'le bunk for the first time in years to fall asleep to the sweet sound of creaking timbers and wind in the rigging. Up again for the late fall daybreak, I joined Mark and Daryl as they vainly watched from the cabin top for any sign of finning dog salmon. The only salmon in the channel were a few thousand Atlantics in a fish farm anchored just outside the fishing boundary and blocking what had been one of Herb Assu's favourite spots for setting his net. The tree in the cedar grove that I once tied the net to was now securing the mooring line for a fish farm net-cage. As a late fall sunrise burned the mist off the low peaks on East and West Thurlow Islands, Mark took the *Splendour* back out into Johnstone Strait.

Dwarfed by the mountains, other seiners were holding position along the Vancouver Island shore. Like his dad, Mark ran past them a few miles to The Slide, a mile below Harry Moon's Point. Harry Moon was a well-known fisherman who fished at what the marine charts call Camp Point so consistently that, among the fishing fleet, the point bears his name. "The Slide" might similarly have been known as Herb Assu's Slide as Herb was one of the few who understood how to set a 220-fathom net in the quick-shifting tides that characterize the spot. Herb's brothers-in-law also learned these tides. This Sunday, as Mark anchored at The Slide, his maternal uncles Tony (*Western King*), Aubrey (*Royal City*) and Jerry (*Westisle*) Roberts, all held positions nearby as did Mark's cousin Teddy Assu (*Sandy Anne*) who learned the tides from his father Ronnie. This locale is known, sometimes grudgingly, on the coast as a family-controlled spot and this day there were no exceptions. In practice the family territory extends down along the Vancouver Island shore to Humpback Bay, just past Mitzi's Bay, named for Mark and Daryl's mother. Mitzi's baking, done on the oil stove in the tiny galley of the *Departure Bay No. 3* while anchored in that bay to wait for tide, was the envy of the family fleet. Our cook on this 2001 trip was Mark and Daryl's older sister Judy, who covered the baking requirements with some of her mom's fresh buns and a couple of banana loaves baked at home and sent along.

One of the most significant changes since I was a

Daryl and Mark Assu's uncle Tony Roberts with the *Western King*, built by Bensons in 1964, on a set at Humpback Bay in Johnstone Strait.

fisherman is the division of the coast into two seine areas, each requiring a licence. Since no new licences were created when this division was introduced in the 1990s, there are now fewer boats in each area. Though this management change has been devastating for many fishermen who could not afford to buy and stack a second licence on their boat, it could benefit skippers like Mark and Daryl. Like their father, their grandfather Dan Assu and their great-grandfather Chief Billy Assu, they are Johnstone Strait fishermen with little desire to fish "up north" and still less desire to have others fishing in their territory.

By 9:30 am the *Splendour* was at anchor in the kelp beds off The Slide. As their father did when he anchored at this spot ahead of the Sunday night opening, the brothers sat up on the flying bridge with the crew and watched the tides for sign of salmon. "They usually show before and after low water," said Mark. As he predicted, at 10:30 a school of chum salmon began finning just outside the boat along a line where the flood had started on the beach

Assu and Roberts family boats waiting for their turn to set.

Drumming back the *Splendour*'s net.

and pushed against the last of the ebb. Just offshore this created a riptide that seemed to stimulate the fish or simply gave the promise of a fair tide to carry them along toward their home rivers. As we did back in the 1960s, the crew lamented that the opening was not until the next day. "Looks like a couple of thousand in that school," observed Daryl. "I hope there are some more this time tomorrow."

Sunday evening, six o'clock, the traditional opening time, passed quietly with the crew working on deck to ready some of the newly introduced technology. The brailer, that was largely replaced by a hydraulically operated ramp stern in the 1980s, made its return in the '90s to facilitate the release of non-targeted salmon species. The crew placed a sorting box on top of the hatch combing that allows the non-targeted species (this week that included coho, chinook, steelhead and sockeye) to be removed to the recovery box for later return to the sea. The

Getting ready to brail dog salmon on the *Splendour*.

recovery box looks like a half-sized tote with an electric pump to circulate sea water.

A smaller, secondary boom is also an addition since I last worked on a seiner. It evolved from a similar boom used to handle the fish pump on herring, but here it will be used to hold up a 12-foot section of aluminum pipe to which the cork line is tied while brailing. This eliminates the need to tie the skiff to the cork line, making one more crew member available on deck to sort fish. I had heard that in the years since I seined, most fishermen have become accustomed to taking maximum advantage of the increasingly limited openings. If it was a twenty-four-hour opening they fished right through the night. On a twelve-hour opening, they kept drift setting even when the tide wasn't what had been considered "right" in the old days. But the skippers fishing The Slide were tide-fishermen who use their knowledge of the tricky currents to make a limited number of sets at optimum conditions.

This presents a problem when an opening lasts only twelve hours. "Seems like the DFO sets the openings so that they don't work with the tides," the crew complained.

Monday morning proved the point. When the radio announced that the fishery was open at 7 am, the *Splendour* stayed at anchor for three hours with the ebb tide backing up at two or three knots just off the shore. Mark had come out early to hold the position that he knew would give him an opportunity for the combination on which he, like his father, thrived: the even push of a beginning flood and a school of finners going into the net. With the anchor finally up, he circled the boat in toward the beach, watching the effect of the tide along the shore. He watched as the rip that indicated the change moved offshore. Just before 11, when he saw and sensed that things were suitable for the fish to begin moving, he made his final pass at the shore and signalled for the skiff to be let go so that the net could begin paying out.

Brad Roberts, who once skippered his own seine boat, ran the skiff's outboard while Tarzan Scow ran up the beach to tie the end of the net to the peg. Farther up the shore, the uncles Tony and Jerry were letting their nets go. With the net out Mark towed easily with the last of the ebb holding the boat end of the net up against the first push of the flood along the beach. Judy, who had come up from the galley to watch the first set, said, "Tarzan isn't dancing a jig on the beach yet so he must not see any jumpers going in." But Tarzan did see finners going in even though on the boat we saw nothing in the rising chop of a twenty-mile-per-hour westerly wind. With the end drummed back onto the boat and the rings up, it took only minutes to have all of the net up to the bunt drummed back aboard. The drying up was quick and efficient; with the cork-line bar hooked in place and the brailer on the single-fall, the crew worked as a smooth team to dip and dump sixteen brailers for nearly one thousand big dog salmon.

A big man like his father, Mark moves around the deck with the same calm demeanour that instilled confidence in deckhands a generation ago. The same big hands and stubby fingers danced knots from tangles and the powerful arms pulled reluctant lines. With the fish and gear

Except for the addition of the outboard motor, the jobs of the skiff man and the tie-up man haven't changed in decades.

safely aboard, Mark turned the boat down along the shore to look for his next set. With the flood tide moving along the shore as well, he ran down past Mitzi's Bay and picked a tree with a rope-strap on it for Tarzan's next tie-up spot. The skiff dropped from the stern of the speeding boat and Tarzan dropped the sea anchor into the water to start the net off the drum. In my fishing time back in the '60s and

Letting go with the skiff man making for shore and the net drumming out.

'70s, the skiffs were mostly made of wood and were always rowed to the beach with the skiff man standing to push on a pair of nine-foot oars. Now the aluminum skiffs were powered through the kelp and into the beach with an outboard motor. While diesel power skiffs are still illegal for inside waters like these, outboard motors have become the norm.

In the past thirty years, horsepower ratings on many of the seine boats have increased from the 100 to 200 hp range to well over 500 on some vessels. With heavier gear this allows for towing against stronger tides. It has also necessitated the drilling of holes in the Vancouver Island bedrock into which the tie-up man places a steel post for tie-up, as most of the old wooden posts and tree stumps have been towed out of the ground by the powerful boats. Mark is still working with 245 hp. This tide is relatively small, but Mark still sets his net so that the boat begins to tow well up from the beach end of the net. Like his dad, he picks a couple of marker trees up on the steep hillside above the tie-up and watches as his boat, at full throttle, is pushed back in relation to the shore. At the last possible moment, Mark makes the turn into the beach end and, with the boat listing to port, he signals for the end to be let go. Another innovation since I fished takes over when the running line attached to the large drum of the pursing winch begins to wind the beach end back to the boat. It is a line that runs out with the net but is attached only at the beach end. As the boat tows, the slack is taken out of the running line making a direct connection from boat to beach. With the freshening wind, now at 25 miles-per-hour, this makes a skiff man's life a good bit less stressful.

Other things never change. As the tide carries the *Splendour* down past Humpback Bay, a rocky point that juts out and into the stream threatens to impale the boat

The skiff men returning the end to the boat.

broadside. Mark is at the top controls using another more recent innovation, the bow thruster, to keep the stern of the boat pointing in the direction from which the net is being recovered. But the wind is pushing the boat faster than the tide, and web that has billowed under the boat's keel is sucked into the bow thruster. With typically calm practicality, Mark climbs down into the skiff and fires up the 20 hp outboard that he leaves running and tied to the side of the boat. It is just enough power to move the boat with the flow of the water out to clear the point. Mark's dad had neither a bow thruster nor an outboard-powered skiff, but by directing us to pull corks and waiting until his propeller was clear, he could usually kick the boat around the same point. Only when all else fails will a proud skipper ask for a little tow from a cousin or brother fishing nearby. That doesn't change. The second set of the day yielded a couple of brailers of dogs. Minutes after the last fish was aboard, Mark had cranked the Jimmy up to full revs and was headed back against the flood tide and a 35-mile-per-hour westerly that was now whitecapping the blue waters.

But fishing is about competition, and one of Mark's uncles with a bigger, more powerful boat passed him and got ahead of him in the line to set at Humpback Bay. Meanwhile another uncle was getting towed off the point by a cousin while still another cousin's boat lay at anchor in the tight confines of Humpback Bay with web wrapped in his wheel. This is a tough spot to fish at the best of times and when you are working with a crew that has had only three days of fishing all season in strong winds, trouble is always nearby. And that is the great difference between 2001 and my fishing days. We had regular two- and three-day fishing weeks with occasional four-day weeks. There might be a ten-day tie-up from time to time, but we had enough fishing to attract and hold a crew that learned to work together as a team. When I stepped back on Mark Assu's boat in 2001, the southern seiners had had three days of fishing: one in August for sockeye, another for pinks and this one on the first of October for chum. A couple more chum days followed later in October and November. In the years since 2001, crewing has only become more challenging.

By the time the gear was back aboard from the third set, it was nearly 4 pm. Mark and Daryl decided to take a look back over in Blind Channel. The westerly was pushing hard directly along the channel's beach when they arrived about 5 pm. It made it difficult to set farther out but the pens full of Atlantic salmon still blocked the more favourable tie-up spots inside the channel that my old skipper Herb Assu would have used in these conditions. Mark decided to set in spite of the wind. It was a bit tricky as they would tow the opposite way to what they had been doing and purse the net on the starboard rather than port side. The set went well in spite of the wind. I recalled that in the early 1960s we would set in that wind with a power block up on the end of the boom; it created a sail of web that would drive us well up into the corner now occupied by the fish farm. With the gear back aboard, there probably wasn't time to make another set before the 7 pm closing and there certainly weren't enough fish to risk the fish cop's wrath. The tide had turned to ebb so we wouldn't get through Seymour Narrows until much later.

Mark ran the boat the few miles on into Blind Channel store. When I knew Blind Channel in the 1960s and '70s it was a store and post office serving commercial fishermen and loggers. But now, with the logging and fishing dramatically reduced, it thrives as a tourism destination with a classy restaurant and finger slips for yachts. A

Daryl and Mark moving a herring seine net from the *Western Girl* to Daryl's *Adriatic Star* in preparation for the trip to Kitgatla. The crew includes Elvis Chickite on the lead line and Tarzan Scow on the cork line.

fish-dressing station on the dock bears testimony to what happens to many of those coho and springs that we had spent the day carefully releasing. The *Splendour* moored to the dock looked a little out of place, but not apologetic, in the grand surroundings. Later, with dinner and gear stowed, we ran the three hours down through the Narrows under the huge Dog Salmon Returning Moon. Below the Narrows we tied up alongside a seiner at a small processing plant that had agreed to buy Mark's fish. Most of the other boats, flying the red and white Canadian Fishing Company (Canfisco) flag, stopped on the way to deliver their fish to the only company packer in the area. In my fishing days, three big companies—Nelson Brothers, BC Packers and Canfisco—and often one or two smaller companies all had packers on the grounds. By 2001, they had each been purchased in turn until all their fleets were ultimately owned by Jimmy Pattison.

Another difference for Mark is his tenuous fishing rights. When his fishing contract was sold from one company to another, he decided that the only way that he could escape corporate control was to sell his salmon licence to the government-sponsored buyback program and pay off his company debt. Daryl, who owns the seiner *Adriatic Star* did the same. The result was that the two brothers owned two boats and only a single herring seine licence. As Natives, they have the opportunity to apply for the temporary use of a southern salmon seine licence from their tribal council. Mark had won this in a lottery-style selection for the past two years, but the future remained uncertain. Daryl's boat remained tied up during the salmon season. It was 3 AM when they moored at the plant. At six their turn to deliver came up and the dock man called down to Mark and Daryl, "I've got a farm fish boat coming in here to deliver at 7 so you have to go and

A pair of southbound salmon seiners buck an ebb tide through Seymour Narrows.

drop your pick until he is finished and then we will take you."

"We aren't going to be bumped by any damn farm fish," replied the independent brothers. They ran their boat over to Campbell River and called another buyer who sent a truck to take their 18,000 pounds of thirteen-pound chums at fifty cents per pound. With the fish delivered and the boat scrubbed down, we retired to one of the family's houses for a salmon dinner and fish talk with crews from other family boats. Around the table were assembled a couple of hundred years of combined fishing experience. The challenges of fish farm competition, the loss and corporate control of licences, the twelve-hour fishing week and the restrictions for by-catch avoidance were all fodder for discussion and anger. But it was Mitzi Assu, my old skipper's wife and the now retired cook who baked those great pies in the bay that bears her name, who summed up the current state of the fishery most clearly. "It's just not fair," she said.

16

The Arnets
Three Generations of Changes

"George's father didn't do any of the fishing that we did, and our son Russ will have to change with the times," explained Sue Arnet in the couple's beautiful house set high up on the North Vancouver mountainside. Sue went on to explain, "The herring is why we are here, but it is disappointing to see it go." George and Sue had returned two days before from a golfing holiday in Palm Springs. Their forty-year-old son Russ had just come down from his Prince Rupert home to go out to the Gulf of Georgia for the opening of the 2009 herring season. Russ was running his dad's boat, the *Attu*, with the licences from his own boat, the *Caamano Sound*, stacked along with an additional licence from the small fish buyer to whom he would deliver the herring. But change was in the air. Most fishermen I had spoken to in the weeks leading up to the March 2009 roe herring opening maintained that the fish were becoming smaller and that the Gulf should have been closed to give the fish a chance to come back.

Most of the other areas of the coast had already been closed and older fishermen were recalling the demise of the herring fishery that happened after over-fishing in the 1960s reduction herring fishery. I recalled taking part in the reduction fishery and hearing the older fishermen accurately predict dire results if we kept using street lights on our boats to pit-lamp the herring in the Gulf Islands.

George's grandfather had emigrated from Norway in about 1888. He came to BC in 1890 to try the fishing on the Fraser River and eventually settled on the west coast of Vancouver Island near Tofino. George's father Edgar, born in 1899, went out on one of the early halibut boats as a sixteen-year-old. It was the kind of hard-driving heavy-weather fishery he had heard about from his dad's North Atlantic homeland. In 1925 he built the halibut longliner

George and Sue Arnet at their North Vancouver home, March 2009. George is the long-time owner-operator of the *Attu*.

George Arnet with the *Attu*. ARNET FAMILY PHOTO

Egon Pedersen holds a big halibut on the *Attu* about 1992. ARNET FAMILY PHOTO

Cape Beale in a shipyard on Vancouver's False Creek. In 1929 he and the *Cape Beale* earned a place in fishing history when he jury-rigged a set of paddle wheels to come in from the Gulf of Alaska when his tail shaft broke.

Born in Prince Rupert in 1936, George moved with the family to Vancouver where he grew up around the smell of yellow cedar in local boatyards with a clear understanding that there could be a good life in fishing. By the time he and his dad had the steel-and-aluminum combination boat *Attu* built in 1959, Edgar was getting ready to pass the wheel to his son. The *Attu* was built, in part, with the insurance money from Edgar's halibut boat *Kodiak,* which had sunk when a plank let go with a full load of herring that they were packing from the west coast. It was common practice in those days for halibut boats to find off-season work packing herring and salmon.

George wanted to try seining, so the *Attu* was built with that capability but it was named for the island in the

Aleutians in honour of Edgar's halibut fishing grounds. Edgar and George continued to take the *Attu* out to the Bering Sea and the Gulf of Alaska to longline halibut in April, May and June. Most years, they could fit in one three-week trip to the Bering Sea in April followed by two shorter trips to the Gulf of Alaska. The turning point for that fishery came when the recognition of two-hundred-mile Exclusive Economic Zones in 1982 ended George's Gulf of Alaska halibut fishing. By that time, the herring roe fishery was earning Canadian fishermen big money and the salmon fishery was doing well. George and Sue had a young family and life was good. Neither of them missed the long voyages, with the long separations, to Alaskan waters and there was enough money to go around.

Some years earlier, George had bought out a share that the Canadian Fishing Company (Canfisco) had in the *Attu* so he could pay his father a proper boat share for his ownership. They had originally allied themselves with Canfisco in order to fish herring in the reduction fishery when they built the boat in 1959. George explains that the four companies that had reduction plants—ABC, Nelson Brothers, BC Packers and Canfisco—controlled the seventy-odd licences for that fishery. After gaining independence, the father-son team had been fishing for Andy Tulloch's Tulloch Western Fisheries until his company went bankrupt in the 1960s. Then they joined the newly formed Central Native Fisherman's Cooperative in part because respected managers Eric Kramer and Leif Nordal were there. "My dad used to say, 'There isn't a fishing company on the coast that hasn't gone broke,'" explained George. Central Native followed the pattern. When Kramer went to work for British Columbia Packers (BCP), one of the two big companies that still existed on the coast, George and Russ followed him there.

The *Attu* at anchor in 2004. *ARNET FAMILY PHOTO*

But conflict arose when the Weston family's BCP decided to wind down the century-old firm's role in the BC fishery. Having sold off much of its real estate, the firm sold its fishing operations in April of 1999 to another century-old firm, the Canadian Fishing Company. The sale to Jim Pattison's Canfisco included not only BCP's own fishing boats and salmon fishing licences, but also their fishing agreements or contracts with independent vessel owners. In announcing the sale, a form letter to their fishermen signed by BCP Senior Vice-President Garnet Jones stated, "The processing operations in Alaska and BC, as well as the wholly owned company vessels, are being sold to Canadian Fishing Company." Touching the nub of the controversy, Jones continued, "Current or long-term fisherman accounts will be transferred to Canadian Fishing Company when the sale is completed . . ."

Most fishermen went over to the new buyer. But at least one fisherman I spoke to at the time sold his salmon licence to the government buyback plan that was using tax dollars to reduce the fleet and increase the viability of the industry. He apparently felt that he had no other option if he wanted to remain independent. Four fishermen known as strong producers—brothers Gordie and John Wasden along with father-and-son team George and Russ Arnet—decided to fight the deal. They felt that they had agreed to sell their fish to BCP in return for prices, services and other important considerations. The considerations included the sale to them of a second salmon licence when the coast was divided into two areas under the Mifflin Plan. The companies, anxious to retire a number of older wooden boats, were eager to stack licences on modern boats with productive owner-operators, so they sold second licences to selected fisherman at a favourable price. When BCP sold to Canfisco, the two big packing firms maintained that the balance of multi-year fishing contracts signed by fishermen with BCP had to be honoured by the fishermen with Canfisco. They argued that the fishermen were contractually obligated to sell their fish to BCP's successor. The four fishermen approached lawyer Ray Pollard for advice. He agreed that their contracts could not be sold—in legal terms, assigned—without their prior agreement. He wrote a letter for them explaining this to Canfisco and BCP before the 1999 salmon season.

The four fished an early opening in the 1999 season in the northern area and sold their catch of pink salmon to independent buyers. Lawyers on behalf of a numbered company the fishermen assumed to be an arm of Canfisco immediately requested and were granted arrest orders on the vessels. John Wasden of Alert Bay recalls events leading up to the arrest. "They were taking pictures of us delivering our fish," he says. "I didn't hesitate for a moment. I knew that what I was doing was right. After they arrested my boat, the *Western Eagle*, I made one delivery to them and then just tied my boat up. My brother had already tied his boat up rather than fish . . . against his will."

The Arnets petitioned the Supreme Court to release their two vessels, the *Attu* and the *Caamano Sound*, pending settlement at arbitration. The court noted in its finding that "BC Packers Ltd. (BCP), in June of 1997, entered into an Agreement with the defendant for the sale of a salmon fishing licence." The court then noted that this required the fishermen to sell their fish to BCP, and gave the company recourse through seizure of the vessels if the fishermen didn't deliver their fish to BCP. At the same time, Judge Taylor noted that the information provided to the court by the numbered company "reveals nothing as to who the principals of the plaintiff are, its assets or

Russ Arnet's *Caamano Sound* was one of the boats chained to the dock in the contractual dispute.

its ability to perform any of BCP's obligations under the contracts. The defendant [George Arnet] exhibits concern in this regard and rightly so. The defendant's position is that the Fishing Agreement is a personal contract not capable of assignment [to another fishing company]." Taylor observed of the agreement, "It is obviously a contractual arrangement indicating mutual trust and reliance by the parties . . ." Taylor found that the circumstances of the arrest had "disclosed an abuse of the process of the court . . ."

Staking out a middle ground, he left the arrest warrant in place while granting the Arnets the use of their boats on the condition that they delivered to the numbered company until arbitration could settle the matter, at which time any damages for lost revenue would be paid. In the fall of 1999, the lawyers for the numbered company agreed to go to arbitration on the issue. While other fishermen warned of the deep pockets of the corporations, the four never wavered in their commitment to seek justice and keep on fishing. The seven days of arbitration hearings in Judge Taylor's court revealed some of the methods by which the largest companies in the BC fishing industry use federally granted fishing licences to pursue their interests. The companies were reluctant to sell licences to fishermen for a variety of reasons, including because they sought to control production from licences. It emerged that BC Packers, like other processors, sold a licence only when the production from that licence could be secured over a number of years through a long-term fishing agreement. Because BCP had done business in Alaska through its Nelbro subsidiary, BCP officials were well aware that Alaskan packing companies don't have the advantage of owning fishing licences—called "permits" in Alaska. Alaska restricts permit ownership to owner-operator fishermen. Nelbro was sold to Canfisco at the same time as BC Packers and now does business in Alaska under the name Alaska General Seafoods, which also includes the former Kanaway Seafood, and Alaska General Processors.

In the case of Russ Arnet, whose grandfather Edgar helped develop the BC halibut fishery, BCP had allowed him the use of a halibut licence and quota that they owned. Another deal revealed in the legal dispute was BCP's sale of a restricted-ownership Native herring licence to John Wasden's First Nations wife Theresa in return for a delivery commitment from John. An issue in the arbitration was the difference in what the numbered company claimed to be the market value of the licences and the lower amounts at which they had been sold to the fishermen. The Wasdens, who are Native, and the Arnets, of Norwegian descent, had been members of the Central Native Fisherman's Cooperative before it folded. Its respected manager, Eric Kramer, took employment with BCP. The Wasdens and Arnets followed him. It was that personal nature of the contract that was a deciding factor in the arbitrator finding that the contract could not be assigned to the numbered company by BCP. Taylor's fifty-page finding also noted that there is no specific provision for the fishing agreements to be assigned or transferred:

> The assignments were part of a process by which BC Packers removed itself from the fishing business so as no longer to be able to offer the fishermen services or benefits, nor provide a market for their fish, price their catches or establish a minimum price for them. When BC Packers gave notice of its withdrawal from the business and its assignment of the contracts to the numbered company, and the solicitor for the fishermen indicated that they would treat the contracts as at an end and they plainly did so by their conduct, it seems to me that the requirements for a duly "accepted" repudiation [of the fishing agreements] were met.

In his final ruling in February 2000, Taylor went on to say that the fishermen owed nothing to BCP for the licences that had been sold at subsidized prices. Furthermore, they could pursue their claims against the numbered company for losses resulting from the arrest of the vessels and subsequent sale of their fish to the company at reduced prices. The numbered company appealed the decision, but in the end the appeal did not proceed. Contacted in 2000, Don McLeod, Canfisco's Vice-President for Production at the time, said of the arbitrator's finding, "I don't know that it means a great deal. You learn from your experience. In future people will structure [agreements] a little more clearly."

In 2009 John and Theresa Wasden sold their big

The *Attu*, in about 2001, with a 300-ton herring set. The *Sandy Anne* is on the cork line. ARNET FAMILY PHOTO

aluminum seiner with its herring net and power skiff to a California fisherman who will fish sardines and squid. BC boats are much in demand in the healthier fisheries of California and Alaska. The Wasdens, whose son had taken work in the tugboat industry, sold their two salmon licences to a government buyback program that would likely turn them over to a Native community. John remained much in demand as a skipper, so was asked to fish the aluminum *Pacific Horizon* for the estate of a fisherman who had recently died. Wasden still owned his herring licences and took them onto the same boat for the 2009 seine fishery.

In early April 2009, I joined John and Theresa to watch their granddaughter perform spectacular spins and jumps as a figure skater in Campbell River. John, who was fifty-seven at the time, explained that his whole life has been that of a fisherman. "But when they brought the individual quotas to the herring fishery I was in an area where

In 2009, Russ Arnet was concerned about the future of the salmon and herring fishery and had bought a crab licence. The Hecate Strait crab fishery is now threatened by wind farms. Whatever happens, Russ, like his father and grandfathers, plans to stay in the industry.

Byron Wright was fishing. I heard him on the phone with another boat. They had both found the same school of fish and Byron asked the other guy if he wanted them! That's when I knew that it was over." The point of John's story is that the late Byron Wright was notoriously aggressive as a competitive fisherman and had never been known to give space to another fisherman, but with individual quotas there was no competition. While this system has distinct advantages for quality and stock management, it had removed the competitive nature of the fishery with which John had grown up.

Wasden sold his boat because he felt that there was little place for him as an independent owner-operator. "At first it felt like a big relief when I sold the boat. I honestly don't miss it because they drove it out of me. But when it gets down to a point where a bona fide, coastal Native fisherman has no place in the industry it is a sad day."

In 2009, the Wasdens and Arnets operated in a dramatically changed fishing industry. While much of the old relationship-based way of doing things remains, there is a much greater reliance on accountants and lawyers in the making of licence and leasing agreements. While there is no doubt that the old ways were paternalistic and even exploitive, so long as the resource was available to the best fisherman, the field—if not the sea—was level. As limited entry and transferable quotas have increasingly become the norm, the privilege to fish has increasingly taken on the character of a capital investment and to some extent a property right. A few days after the 2009 Gulf herring opening, Russ Arnet was doing some minor repairs on the *Attu* as he prepared to go north to fish a herring opening at Kitgatla. "It used to be a big deal to be an independent fisherman." He explained that the value of the licences to the companies, even small independent companies, is so great that independent fishermen bring their licences to the buyer who matches them with additional licences to generate enough quota to make the fishery viable. So even "independence" is limited and controlled by the fish buyer.

After Russ invested in two licences so he could fish both north and south areas, the DFO called short openings in both areas on the same day—effectively disallowing him the opportunity to fish both. "Even though I bought into the DFO plan to reduce the fleet by purchasing a licence from a boat which would then be retired, I didn't get to fish it. I liked salmon fishing, but after that I sold the licences." So, just as his dad moved from halibut to salmon and herring, Russ moved from salmon and herring seining to fishing crab with traps called pots. He sold his two salmon licences and borrowed extensively to buy a crab boat with a licence to fish eight hundred pots. The mortgage to make this purchase was daunting, and even the legal fees for the transfer were significant. "The banks say that the licence has no value for the mortgage but if you take off the licence they won't give you any value for the boat."

Russ is optimistic about the crab fishery but already he is concerned as he sees plans for a wind farm beginning to encroach on the crab grounds in Hecate Strait. "I won't push my son to be a fisherman but if he chooses fishing I hope that it can have enough diversity for him to make it." A fisherman in today's industry who does not want to become the equivalent of a grocery bagger in a vertically integrated food empire will need to be sharp to maintain the independence that was the hallmark of fishermen like the Wasdens and the Arnets. "And," concludes Russ, "keep the boat working."

Billy Griffith and the Kaislas

Still Fishing Prawns

The late Iris Griffith once rowed with her mother the 500 miles from Haida Gwaii to southern Vancouver Island in order to go back to school while her father, Vic Hill, stayed up north for the balance of the salmon trolling season. Iris's husband, Billy Griffith, still fishing at seventy-five years of age, is made of equally solid stuff. Billy carries with him a well-earned reputation as a successful fisherman, but beyond that he is known on the BC coast as an absolutely and consistently honest man. He is proud of the successes that he and his crew on the purse seiner *Tzoonie River* have had releasing salmon species from endangered stocks. While these releases were mandated in some openings, not all fishermen, in the rush of fishing, were as conscientious about their care in the release of fish. But Billy said that his crew even voluntarily released big spring salmon around the Gordon Group off northern

Vancouver Island one summer as they felt there were not enough getting through to the spawning grounds.

It had been at least a decade since my last visit when I stopped by Billy's waterfront acreage near Egmont in January of 2010. Not much had changed. The blackberry and the salmonberry still vied with each other to embrace dozens of retired and semi-retired trucks, boats, marine engines and even a steam donkey that Billy has collected over the years. "And I have a bill of sale for each one," he declared proudly. As we stood beside a shed made of hand-split cedar shakes, Billy told of his attempts at quarrying slate from up Jervis Inlet. Such projects were a tradition of BC fishermen. Billy has worked at logging, clam-digging and any other source of cash to tide him over between fishing seasons. But these opportunities are getting to be slim pickings.

I wanted to know how Billy was doing with his prawn licence. Born and raised in Egmont at the mouth of both

Billy Griffith's boat shed at Egmont. Few fishermen can still afford waterfront property.

Jervis and Sechelt Inlets, Billy has lived his whole life around prawns and prawn fishing. Today, with no freezers on the *Tzoonie River*, he has his prawns trucked live to Vancouver. But he chooses not to go out every day. This casual approach, which is in keeping with life at Egmont, doesn't work so well for a buyer with firm orders from the big city. "When they made the prawn fishery a limited entry, I couldn't get a licence because they [DFO] said that I had not put in enough catch on the two qualifying years," he said as we reflected on his decades of fishing the inlets and those decades that his father before him had fished those waters. Billy explained that he hadn't fished enough in those years as he was working full-time as a volunteer on salmon stream enhancement projects.

Like so many stories on the coast, this one reflects the inequities of a system that allows a wealthy investor to buy a licence and then lease it to a genuine fisherman. At the same time, this system can deny a licence to a fisherman and his wife who have generations of fishing behind them and make a commitment to the future of their resource. Billy protested the unjust decision. As a result, "I was grandfathered in," he explained. "That means that I can fish prawns but I can't leave the licence to my daughter." Billy recalled how one of the great processors and champions of the BC coastal fishery, the late Don Cruikshank, wrote a report on licensing in the BC fishery in 1991. Commissioned by several commercial fishing associations, it was largely ignored by the federal government. "Don Cruikshank said that the holder of the fishing licence should be a fisherman on the boat," Billy said. The report predicted that transferable quota shares would result in the privatization of the resource. Cruikshank was also concerned that such owners and their urban corporations had little affinity for the people or environment of

Billy Griffith in his yard at Egmont.

the coast. Over the past two decades, Cruikshank's concerns have been proven entirely justified.

For many young people, the emotion of loss of a great publicly accessible fishery is acceptable. For those of us who recall and cherish the idea of a common property resource, the privatization that took place in spite of Cruikshank's report has done much to devastate the spirit and community of BC's fishermen. "Now we have lost our honour as fishermen," Billy lamented. "This year on the Nass and Skeena we were required to brail our salmon so that we could safely release not-targeted species. But there were at least a dozen boats using their ramps out of about forty-five boats fishing."

We recalled how in the 1960s local fish wardens were hired to stay in cabins at the mouths of strategic places. They were there to police but their role was primarily preventive, "I remember one coming over to visit us when we tied up our net on the beach in Gunboat Pass," I recalled.

"Yes and he would not have been wearing a big hand-gun like the one that came on my boat this summer," said Billy, adding that there are now far fewer government people on the fishing grounds. "They have added office staff in Vancouver and Ottawa but reduced the people on the water." Whenever a couple of fishermen get together on a beach, even in the rain among the salmonberries, there will be a certain amount of bitching. It is part of the ritual, just like saying, "Next year will be better." But Billy is a fine and sensitive human being. He speaks from a deep and abiding concern for the resource and the lives of the fishermen who have worked it—and he is discouraged.

Billy isn't the only one who is discouraged. "I am getting tired of the prawn fishery. We only get about fifty-eight days per year. To expand I would need to buy halibut quota but that would take twenty years to pay for,"

Kali Kaisla explained as we sat on his boat, the *Eskimo*, at the Sooke government dock. Kali and his mom Ruth had put the boat and its spot prawn licence up for sale in 2003 when Kali's dad Jim died, but there were no takers. This lack of interest was probably because prawn licences are limited entry and operators must fish a limited number of traps over a limited number of days. An individual transferable quota would allow fishermen with boats already carrying several licences to purchase and stack additional prawn licences. It is the individual transferable quota that drives prices up, as it attracts non-fishing speculators who want to lease the licence and guarantee their own supply of product that they can then resell at higher prices.

Kali thinks that a quota prawn fishery would benefit him, but doesn't hold much interest in the transferable part. "Most of the licences are now fished by their owners," he explained. "This is quite a family-oriented fishery. Each licence is allowed 300 pots. You can stack two licences on a boat but then you would be allowed only 450 pots and most boats aren't large enough for that many anyway." In the 2007 prawn season, 224 boats that delivered 2,500 tons of prawns fished the coast's total of 258 licences. These numbers indicate very little stacking and confirm Kali's assertion that it is largely a family fishery. It is also well regulated, with a season running from May to July. By mid-May, most females have spawned in the cold deep inlets along the coast. Throughout the season, technicians operating from their own boats are contracted by the Department of Fisheries and Oceans to board the prawn-fishing boats while they are hauling their traps. They sample the traps to make sure that the ratio of males to females is such that there will be enough females for a healthy spawn in the following year. It is possible

Kali Kaisla in the galley-wheelhouse of the *Eskimo*.

to close small areas where the ratio is not adequate and fishermen can move to other areas.

To slow the fishery and allow for more precise management, the crews are permitted to work only from 7 am to 7 pm. Each licence is limited to only three hundred traps, called pots, and regulations stipulate that they can be hauled only once a day. This works well for Kali and his crew. "We usually finish hauling by 3 or 4 in the afternoon. We have a live tank where we keep a whole day's catch alive while we are hauling. When we are fishing in an inlet with a large river we have to put a hose over the side to suck up cold salty water from ten fathoms down to keep the prawns alive because the surface water is too fresh." Prawns are one of the few fisheries where the fisherman can package for consumers right on board. Kali and his crew put the prawns in mushroom baskets,

The *Eskimo* moored at the government dock in Sooke in 2009. The *Eskimo* was built by Matsumoto in 1941 just ahead of the *Universe* and his internment.

then dip them in a solution to prevent freezer burn and to make them a bit more docile. They are then sorted and repackaged in four sizes. Finally they are put into a plate freezer on the stern of the *Eskimo*.

It was the *Eskimo* that brought me to meet with Kali at the Sooke fisherman's dock. In the early 1980s I had visited with her builder Sam Matsumoto at his Dollarton shipyard on Burrard Inlet. Sam had told me about building the boat with his uncle in Prince Rupert shortly before being interned as a Japanese-Canadian in World War II. Sam had convinced his uncle to add a bit of beam to the boat to take advantage of the new smaller engines available at the time. The *Eskimo* was registered in 1941 and was followed by the larger *Universe* in 1942, with the builders allowed to stay on the coast a little longer to finish it.

When Sam was finally allowed back to the coast in 1949, he built gillnet boats and then in the early 1950s began building a series of boats including *New Venture*, *Miss Georgina*, *Eva D. II* and *Zodiac Light*. They all trace their lineage to the *Eskimo*. For much of her sixty-eight years, the *Eskimo* was owned by BC Packers. She seined salmon and probably longlined halibut on occasion. Built stout and solid, at only 50 feet, she is small for a seiner—and her 14.1-foot beam is not extravagant by modern standards—but she has about her a no-nonsense look. When Kali's dad bought her in 1995, the purchase brought together two strong and deep-seated BC maritime communities. Kali's great-grandfather, Karlo Kaisla, was one of the pioneer Finnish immigrants to the utopian community of Sointula on Malcolm Island. An artesian spring near Pulteney Point on the island still bears his name because he drowned off the same point at about thirty-five years of age. Kali's grandfather, for whom he is named, was born in Sointula but homesteaded and settled at Bella Bella between the BC Packers store and Shearwater. The house he lived in still stands there. The senior Kali trolled from there with a boat called the *Bethune* before building the *Gail Linda*.

This former boat was passed down to Kali's father, who was born on the central coast in 1943. At one point the grandfather moved the whole family, including Jim, his two brothers and four sisters, to Ladner. Jim worked at Menchions Shipyard in Coal Harbour for a time between fishing seasons. But the city was not the place for him and in 1980 he moved the family back north to Bella Bella. Then in 1984 they moved to Bella Coola where Kali graduated from school and his mom still lives. It is also the port where he stores his prawn gear and delivers his catch to reefer trucks for the haul to Cloverdale Cold Storage on the Lower Mainland. The *Bethune* served the family well on both salmon and halibut fishing. But in April 1984, with another crew fishing her, she went down off Rose Spit with a full load of halibut. Kali's dad had been working two boats since having a yard in Sidney build the steel *Kal-Anne* in 1976. Initially the *Kal-Anne* was a longline boat only, but after the *Bethune* sank they put the salmon troll licence onto the *Kal-Anne*. Kali recalls his dad coming back from Alaska with the *Kal-Anne*

Siblings Ira, Diane and Jim Kaisla. Jim was Kali's father. COURTESY OF KAISLA FAMILY

"loaded to the scuppers." Canadian access to Alaskan waters was stopped in the late 1970s by the establishment of exclusive economic zones and the surrender of Canadian fishing rights in those northern waters.

With salmon troll, halibut and rockfish longline and prawn permits all on the *Kal-Anne*, the boat was able to fish most of the year. But in the 1990s, increased regulation began the changes that would leave Kali with only fifty-eight days of work per year. Early in the '90s, a few boats were fishing prawns in between their halibut and salmon seasons. Initially they put little pressure on the stock so the fishery was open for eleven months per year. As increased regulations began to limit the salmon troll fishery, the pressure on the prawn fishery increased. Where the fishery had been primarily exploited as a secondary fishery to halibut and salmon, it gained importance as Japanese markets recognized the quality of BC frozen-at-sea spot prawns. BC Packers, who owned the seiner *Eskimo* for decades, had sold it in 1991 without a licence, allowing them to have an efficient aluminum seiner, the *Western Ace*, built for the licence. An owner had added the prawn licence but was fishing the boat as a day boat.

In order to develop his participation and quality, Jim Kaisla bought the *Eskimo* in 1995 and spent a year and a good bit of money upgrading it to a well-outfitted prawn boat. In addition to insulating the fish hold, he installed the plate freezer and extended the main deck to accommodate the traps and provide additional workspace. To finance all of this, the family's salmon troller *Kal-Anne* was sold to a buyer who insisted on buying the 33,000 pounds of halibut quota with the boat, a decision that has now left the *Eskimo* with not quite enough to earn a proper living. The first year that they fished the refurbished boat was also the year that trap limits were introduced. By 2001,

The *Bethune* on glassy waters.

The *Kal-Anne* tied at the dock.

increased licence utilization and fishing pressure brought the number of days down to seventy-eight with a steady decline to fifty-nine in 2007. Catches remained more or less consistent at 2,000 to 2,500 tons.

The fishery suits Kali as it allows him to visit the many places where he learned with his father. Typically, they fished the central coast inlets between Klemtu in the north down to Rivers and Smith Inlet in the south. "I like to explore all the little holes like Nowish Inlet and up Finlayson Channel to Carter Bay and Sheep Pass and then down Mathieson Channel." Kali carries a book on the boat giving the history of coastal place names and mixes an understanding of the history with an appreciation of the geography, but his greatest focus is on the seabed where the prawns live. "In the Gulf of Georgia where depths are more consistent you can set the same place all season but in the glacier-carved inlets we set our pots close to the beach and only at certain depths. Fifty fathoms is my default depth but I have caught prawns as shallow as ten fathoms and as deep as 120 fathoms," he explained.

When the fishing pressure was not so great, the old-timers had favourite spots. Prawns seemed to like old log dumps and individual fishermen might work primarily in one inlet. "They basically farmed it. They would take only the big ones and throw back the smaller ones, even those of legal size, because they could get them the next year." With other fisheries becoming more restricted, virtually all of the 258 licences are being fished. Kali believes that the best option for these licences will be to get a fixed quota with limits set for specific locations so as not to over-fish one isolated stock. In this manner a fisherman could work slowly and, in some cases, deliver to the fresh or live market.

In the current fishery, at least 90 percent of the prawns

go to Japan where they are highly valued in the sushi market. The tails are served up with just the right translucent colour while the heads are fried and eaten as well. The Kaisla family tried to market the wild-caught spot prawns locally for the same $55 per kilo that they are paid by the Japanese buyers. But the less sophisticated domestic market insists on comparing these tasty morsels from the deep coastal inlets with the frozen product from South East Asian farm ponds. "We tried selling to tourists and locals at the dock in Bella Coola for $55, but we had to lower the price. We do have some regular customers in Williams Lake but most still goes to Japan. Once we even tried renting a reefer truck and taking them to Alberta but that was a disaster. If it wasn't for Japan, we would be out of business." In the spring of 2009, as he prepared for one more season of prawn fishing, Kali submitted the licence for consideration by the ominously named relinquishment program. "If they accept it then I won't be going out," he told me by phone in mid-April.

According to the Department of Fisheries and Oceans, forty-eight "eligibilities" were accepted in February 2008 under the combined Allocation Transfer Program and Pacific Integrated Fisheries Initiative. This list included nineteen salmon gillnets, six salmon seines, nine salmon trollers, three halibut tags with some quota, nine prawn trap permits and some sablefish quota. These licences and quota were purchased or relinquished at a cost of $13.1 million. The intent of having fishermen voluntarily relinquish their permits in return for cash is to convert them to communal commercial licences for First Nations communities. It is a program that has won support from a number of permit holders in the Native and non-Native general fishery and has the potential to correct some historic wrongs. By most reports these communal licences are benefiting the coastal communities by providing some earning opportunities, but more importantly by giving pride of employment to the crews who go out on the boats. However, there are apparently some exceptions. Kali reports that, at least initially, "The prawn licences will remain inactive until the governments can complete treaty negotiations."

At thirty-seven years old as he prepared to travel north to pick up his gear and crew in Bella Coola, Kali reflected on his future. He has all but completed a degree in economics and has added to that a diploma in applied chemistry and microbiology. "The freezers on this boat are the bane of my existence," he said. "I have no licence so can't even buy new fluid for them, but now I am accepted into the refrigeration program at Vancouver Island University." So the fishing will continue to pay for Kali's schooling and a diploma in refrigeration maintenance may allow him to continue working around boats. "My first paying job was in 1984 when I was twelve years old," he added. "Before

The former seiner *Eskimo* is now fitted with freezers for prawn fishing.

that it was slave labour. I have a feeling this boat is not going anywhere but it would be good to have it so that if I ever wanted to go tooling around up the coast I still could."

The prawn fishery allows Kali to visit favourite places on the coast.

18

Russell Sanderson

Youth and Optimism

The BC Fishnet email list (fishnet@island.net) is an excellent forum for discussion among the commercial fishing community on the British Columbia coast. Former fisherman Gary McGill of Port Hardy moderates it with wisdom and somehow manages to keep the fish politics on a constructive path. Most postings deal with the conflicts and stresses of a rapidly changing industry. There is a general tone of lament for the "good old days" while striving to forge a survival strategy for the future.

So the following post stood out for its tone of enthusiastic optimism.

BC FISHNET: Jan. 28, 2007

I am looking for a northern troller to lease for the upcoming salmon season for Springs and/or Coho. I have been trolling salmon as a deckhand, and first mate recently, on the *Harmony Isle* since childhood

Russell with his mother France and father Mike Sanderson in front of the *Harmony Isle*, on shore for bottom work. SANDERSON FAMILY PHOTO

(and gillnetted for two years), on the west coast of the Queen Charlottes and the Johnstone Straits. I have also deckhanded on the *Nordic Rand* for halibut, and nine months on the *Ocean Pearl* for black cod. Recently I have been attending Camosun College Nautical program and have finished my training for my Fishing Master level 4, limited to 100 tons. I am a very hard worker and very dedicated. I am available at any time, please contact me at . . .

So I did contact him. Russell Sanderson told me that he was a twenty-year-old who grew up on his mom and dad's 42-foot troller, the *Harmony Isle*. The past six months "on land to do this school has been torture on me. I am longing for the fresh salty air and the brisk Pacific wind, and I even live for the storms," he replied to my e-mail, adding, "Of course there is always talk of the fishing industry dying out, but I can tell you that I have seen some

Russell Sanderson at the wheel. PHOTO BY KAREEN WESTERN

of the best years we've had over the past few." This was definitely different from the doom and gloom that I had heard from many of my contemporaries who recalled the derby fisheries of the 1970s, '80s and even the early '90s when prices were better and accountants urged fishermen to "build a new boat or else you'll have to give the surplus money to the tax man."

Those were the decades when limited entry and quotas were being introduced to the BC fisheries in one form or another from halibut to salmon and from black cod to herring. The fleet reduction in the salmon fishery together with area licensing saw corporate concentration of licences and reduced fishing areas, for those independents that remained. Many fishermen, dispirited with poor returns and fishing areas, gave up and sold their licences to the government at wildly inflated prices. Others chose to stay home and lease out their quotas and licences. It is in this latter group that Russell sees his opportunity. "I also think that it is a great time for me to start leasing boats

The 1979-built *Harmony Isle* at anchor. PHOTO BY KAREEN WESTERN

as there are more and more restrictions and regulations coming into play these days so that many of the older and stubborn skippers may not want to deal with [them]," he wrote, "which would leave them the option of hiring a young and eager guy like myself . . ."

Perspective is everything. When I began fishing as a teenager in the 1960s, the cost of licences was insignificant and keen young men could save the down payment for a boat with a couple of years working on deck. While Russell has had some good earning years, even spending nine months on the deck of one of BC's higher-producing black cod pot-boats wasn't enough for the down payment on a salmon troller with the recently introduced quota that he wants to pursue. A few days later, over lunch at a Vancouver Island restaurant, Russell filled me in on the family history that had spawned this optimistic young man. There was a history of fishing in the family. His great-grandfather built a 14-foot handliner skiff in Victoria in the 1930s and joined hundreds of Depression-era men and a few women who took to the sea in rowboats to troll with handlines for salmon and halibut. Russell's great-grand-father also did a little island hopping with the rowboat to meet his future wife. Other work provided for Russell's grandfather's early career. But inevitably his grandfather, John Sanderson, found his way back to fishing in 1976 when he was in his mid-thirties. Russell's dad, Michael, who was about fifteen at the time, joined his father as crew on the family salmon troller. In the 1980s they expanded from salmon to include the halibut and black cod fisheries. In time Michael married a woman deckhand from another boat in the fleet. Russell was all but born on the boat and has been fishing with the family since he was an infant.

In the "good old days" Russell would have found a

A classic wooden troller, *Grace,* in Port Hardy.

retiring fisherman who would sell him a boat and for a nominal sum he could have bought a one-year troll licence to fish the whole coast of BC. Today the boat is still affordable for the beginning fisherman with prices ranging from $50,000 to $100,000 for a 40-odd-foot freezer troller, but the licence for that boat will be $150,000 to $200,000—and that is to fish only one of the three areas that the coast was divided into in the late 1990s. This means that costs to fish a three-month season would not be likely to impress a bank manager or any other lender. To make the economics work for such a boat in Russell's grandfather's day, before transferable halibut quota, the fisherman would buy a halibut licence and fish in the Olympic-style fishery with a reasonable expectation of generating a significant supplement to the boat revenue. Similar opportunities existed in a variety of other fisheries such as black cod. Today halibut quota costs in the range of $35 per pound with landed prices under $5 per pound. Once again, these are not very bankable figures.

In early February 2007, Russell was negotiating with the owner of a well-outfitted 45-foot steel freezer troller that carried ATF tags allowing the boat to fish the west coast of the Queen Charlotte Islands with a 2007 chinook quota. Russell would also expect to fish some coho and pinks though there would be a non-retention regulation on sockeye. With a 25,000-pound capacity freezer hold he would only have to deliver to Port Hardy or Prince Rupert two or three times through the season. Leasing rates have been pretty much standardized in the industry with a 60/40-split common. The owner gets 40 percent while the fisherman gets the 60 percent, out of which he pays the operating expenses, including fuel and deckhand costs. A variation splits the gross returns 50/50 after the fuel expenses have been deducted.

Russell did reach an agreement with the owner. "I was so anxious to get out fishing that I agreed to a 55/45 split in favour of the owner with the fuel off the top but the crew and grub paid out of my share," he recalled when I visited him again in March 2009. "The owner came out for the first trip and fishing was great, we were getting 100 springs per day. But then he gave me only a crew share of 15 percent after all expenses, which worked out to about 10 percent after expenses. When I crewed with my dad it was 15 percent before expenses." Even when Russell finally got to run the boat with his own crew, the deal kept changing until he finally did the only thing that he could and took the boat back and walked away. It was a valuable lesson learned for a budding entrepreneurial fisherman.

Russell's love of fishing was by no means diminished by the experience. For the 2008 season he arranged to take out his father's boat, the same *Harmony Isle* that he had grown up on. In addition to a deckhand he took his

Russ Sanderson and Kareen Western at the fisherman's dock in Victoria. The boat in background is a sister ship to the *Harmony Isle*.

When fishing is good, life is good. *PHOTO BY KAREEN WESTERN*

girlfriend Kareen Western. It was Kareen's first time on a fishing boat and the couple had a great season with fish, but also with the joys of the northwest Pacific and the beaches of Haida Gwaii. Kareen, a budding photographer, photographed seascapes and sunsets. They had gone north for the mid-June opening and stayed there until September. With some good deliveries on the books, they took a leisurely cruise down the coast, stopping off to enjoy a soak in the hot springs, explore some beaches and eat fresh seafood. Talking with Russell and Kareen over lunch in Victoria, it was clear that she has been infected with a love of life at sea. The stories serve as a reminder of the way in which the salmon troll fishery has sustained the Sanderson family since Russell's great-grandfather rowed out to the islands off Victoria to meet the young man's great-grandmother.

The couple finished up the 2008 season trolling for chum salmon in Johnstone Strait in October. "You troll really slow for chums," Russell explained. "We were crowded right in with the other boats and at times the tide was pushing us backwards in the tide rip but the fish were biting on the change of the tide." I asked if fishing in that much tide with all the other boats so close was stressful. "No," he replied like a true fisherman. "It was exciting." Unlike the northern area that had a quota on the number of fish caught, the Johnstone Strait fishery was a time quota with two- and three-day openings. The boat would call in to choose the days but once they started fishing they had to keep at it for all three days, except when granted permission to stop because of weather conditions. Russell was able to extend his fishing time by leasing additional days from stay-at-home fishermen. These non-functioning "fishermen" are generally called "armchair fishermen" or "slipper skippers."

Sockeye carefully gutted and stowed. *PHOTO BY KAREEN WESTERN*

Looking to the future, Russell worries about the uncertainty of buying quota in any of the fisheries. "If I had surplus money I would consider it, but only if I found a good deal. But the risk is not in the fishing," he stressed. "It is in how the fishery will be managed. With reductions, the halibut and black cod guys are, in effect, losing quota that they may have paid big money for. It just doesn't seem like a good idea right now."

In March 2009, Russell and Kareen were already looking forward to the salmon season. Russell was hoping to get in a trip as a crew member on a halibut or black cod boat before taking his father's boat out again. When I told him about an electronic bird book that I had just seen, Russell excitedly told me of seeing a rare short-tailed albatross during a black cod trip on the *Ocean Pearl*. "Mike Derry, the skipper, and I are both birdwatchers. We thought that we saw a young one and then the next day

Russ holds up a big "smiley" spring salmon. PHOTO BY KAREEN WESTERN

we saw a mature short-tailed albatross and identified it from books that we had onboard." Like his father, who was president of the Pacific Trollers' Association, Russell is taking a leadership role in the fishery and is already a director for the Area F and H troll fishing areas. At twenty-two, he is the youngest director. He is also planning some additional maritime training to gain his Fishing Master 3 (unlimited) certificate and in the future his Watchkeeping Mate. "I think that I would like to qualify to operate one of those ship-berthing tugs," he said. "Just in case the fishing doesn't work out."

Hutch Hunt

"It's Still There If You Want It"

"There used to be a dog salmon run up that creek," Alfred "Hutch" Hunt told me as we looked out the window of the Harbour View Café at Port Hardy's Glen Lyon Inn. "There are still lots of eagles around here at low tide but not today because the herring are spawning out off the point." This is Hutch's home territory but it is a pretty safe bet that he could give the same analysis of a great deal of the BC coast. In the eight decades he has lived, travelled and fished the coastal inlets and tides, he has earned a reputation as an astute student of their ways. I first heard of Hutch when I was a deckhand on salmon seiners fishing the lower Johnstone Strait in the 1960s and '70s. Older fishermen would joke that Hutch had caught so many fish in the upper strait that there would be very few sockeye coming through to us on their southbound journey.

In the 1980s and '90s I knew Hutch as the owner of always-impressive boats, several of which were named

Katrena Leslie. Hutch wasn't afraid to experiment with new boats, and built with steel, aluminum and fibreglass. The fibreglass *Marc Alan* was the first seine boat that I saw with a trendy island in the galley just like a modern kitchen. Hutch knew how to honour the important role his late wife Gloria performed on the boat. As Hutch turned eighty in 2010, he continued to fish and to provide employment to community members. As hereditary Chief Sesaxolis, he carries on a long tradition of community leadership and support of the people. He lives in a house on the beach at Fort Rupert, just south of one of BC's two major fishing offload sites at Port Hardy. It is built on the same location as his father and grandfather's houses. The Hunt family story is well known on the BC coast. Hutch's great-grandparents were a Hudson's Bay Company factor from England and a high-ranking Tlingit woman from Alaska. When Hunt's great-grandfather was transferred from the north coast to Fort Rupert, his son George Hunt

Alfred "Hutch" Hunt with his pole at home in Fort Rupert.

The Gooldrup-built *Marc Alan* was one of a series of state-of-the-art boats that Hutch has owned over his long career.

married into the Komoyue community and founded a lineage there.

The Hunts' renewed and continuing relation with their Alaskan ancestors is equally significant as their sense of heritage and community in Fort Rupert. Hutch tells of travelling to Ketchikan and talking to an old man outside a gift shop. On hearing Hutch's story, the elder introduced Hutch to some of his Tlingit relatives. In the years since, this connection has proved of value both from a family and economic standpoint. In spring 2009, Hutch's grandson was crewing on a herring boat in southeast Alaska. Other family members have fished back and forth, leading to a number of other First Nations skippers and crews going to work on the healthier Alaskan fisheries. The fact that Canadians have excelled in the use of sonar to find herring has also created a demand for Canadians in the Alaskan herring fleet.

Not only did Hutch track his Alaskan ancestry, he also searched his English great-grandfather's roots. This took him to the archives in Dorchester, Dorset. "We found a whole town full of Hunts back in England," Hutch told me as I looked over his shoulder and out the window of his Fort Rupert home to a graveyard island in the bay. Knowledge of his distant roots has enhanced his appreciation of the tides and fisheries of his birthplace, while allowing him a perspective on their beauty and communal value. This perspective was broadened a few years ago when he bought a small Alaskan purse seiner, the *Patti Lu*, and fished her in southeast Alaska. He was impressed with the government management and communication with fishermen. "I really liked Alaska," he explained. "You deliver your fish and tie up for the night and the next morning you know where the openings will be."

Hutch is the grandson of one of the signatories of a "Douglas treaty" for the Fort Rupert or Komoyue people. While most of BC was not covered by treaty, a few dealt with specific communities. The Fort Rupert example,

Winter fishing and snow on the hills. COURTESY OF HUNT FAMILY

signed in 1851, granted significant amounts of land and the right to continue fishing. The result of this treaty, and the recognition of Native fishing rights at various Canadian court levels, seem to have convinced both the provincial and federal governments that they will be required to give First Nations a significant share of all coastal and river fisheries. The transition of fisheries management is very much a work in progress. For Fort Rupert it has already led to the purchase of a boat for the Band. "They thought that we were all going to die off so they made the reserve lands small," explained Hutch. "But the Douglas treaty also said that we could fish as we did formerly. So when the Band wanted a boat I went looking for what would suit them. I looked around the docks in Steveston and we bought the *Western Sunrise*. They have halibut quota and Band members go out as crew. On the salmon my grandson Marc Peeler has been running the boat. Now the Band is looking for a prawn licence."

Since about 1870 when the first canneries were being established in BC, every decade has seemed to have a new fishery, or new regulation of an established fishery. In the first decade of the twentieth century, canneries like the one in Alert Bay purchased their fish from First Nations who controlled traps or beach seines in the river mouth. In the second decade, the processors introduced purse seiners to get the fish before they reached the rivers and the government made a law that Natives couldn't own them. In the third decade, the government banned "Orientals," primarily Japanese-Canadians, from fishing salmon with seiners—but now allowed Natives to fish them. In the 1930s, the Depression slowed development but Natives increased participation. In the 1940s, the government confiscated all the seine boats that Japanese-Canadians had been using to fish herring. In the 1950s, the United Fishermen and Allied

The *Native Joye* is one of James Walkus's fleet of seiners. *COURTESY OF HUNT FAMILY*

Workers Union, generally with the support of the Native Brotherhood, took advantage of a booming economy to get better prices for fishermen. In the 1960s, the UFAWU's progress allowed more fishermen to build their own boats and to become independent from the processing companies; at the same time over-fishing caused the collapse of the herring stocks. In the 1970s, the Japanese roe herring market was developed and drove up the market for boats. In the 1980s, the herring money and limited entry encouraged older wooden boats to be replaced with larger capacity steel and aluminum boats. In the 1990s, corporate consolidation began to put pressure on the smaller independents and the quota system further raised the cost of an independent getting into the fishery.

The timeline above is an oversimplification of a long and complex history, but it does provide an impression of how variable the fisheries are—and how politicized the power structure has been. Racism has been a favourite tool of control, with fishermen and managers sharing responsibility. With eight decades of life behind him, Hutch has seen much racism and heard a good deal more. His father built his first seine boat in 1925, shortly after the 1922 law change that allowed Natives to have a seine licence:

> Dad built a little 40-by-11-foot seine boat powered by a three-cylinder Easthope. Around 1943, he anchored it out by the island in the bay here and the shackle must have broken. The boat drifted up on the beach and broke a bunch of planks. In 1949 my brother Edwin and I got an old guy to come from Quatsino to help us fix it. Edwin took the engine apart and got it running. We went to Bones Bay and the Canadian Fish manager gave us a little two-and-a-half-strip net to use. In 1951 we put in a Chrysler gas engine and we were the third highest boat in Bones Bay.

Hutch came into the fishery just in time for the big sockeye years of the 1950s. "First we fished the Nimpkish sockeye in June, then we went to the mainland inlets for humps [pink salmon} until the Fraser River sockeye showed up in the Strait." But there were also a great many other changes taking place in the 1950s. The rivers were being logged ruthlessly. In earlier years, the logging had been heavy but now trucks were allowing the companies to drive farther back in the valley, creating roads that slid into sensitive spawning grounds. The Nimpkish sockeye were last fished in the 1960s. In spite of nearly four decades of not being fished, they have still not come back. The Namgis Band is not even allowing a food fishery on the stock. A biologist

Hutch Hunt was skippering Canfisco's *Western Roamer* in August 2006. They were continuing to brail sockeye in Baronet Pass while the packer *Hesquiat* was pumping the fish from their hold. PHOTO BY MICHEL DROUIN

once told me that the continuing lack of shade trees leaves the water too warm for healthy survival of the remaining stock. There has undoubtedly been overfishing on many other stocks as well, and it may take many more years for the rivers to mend from the damage that has been done over the past century.

Over time, Hutch got bigger boats and unlike some "homesteader" fishermen who preferred to fish the bit of coast that they were familiar with, he roamed south and north. In the 1960s and earlier, there was a generally accepted division between Johnstone Strait seiners and the west coast or San Juan boats. "When I started going out to San Juan," Hutch recalled, "I got help from the older high-liners. Charlie Clarke told me, 'You've got to stay right to the end and it will add up.' I went up and fished around Prince Rupert sometimes with just three or four

One of thousands of brailer loads of sockeye that Hutch has caught in his lifetime. *PHOTO BY MICHEL DROUIN*

other boats because if you can't get sockeye you might as well fish humps."

It is this love of the fishery that makes a true fisherman. For men like Hutch, fishing is a competitive adventure but also a cooperative effort where other fishermen are helped and knowledge is shared. For all the fishermen who were involved in the Central Native Fisherman's Cooperative, the stories of building a fishing company that was truly supportive of fishermen remain special. There were a lot of good fishermen involved like George Arnet, John Wasden, and Vivian and Bill Wilson. "Vivian and I ran the herring," Hutch recalls fondly. The failure of that venture to gain the financing to keep going, in spite of the fact that they had top fishermen and a warehouse full of finished product, continues to perplex the people who were there.

The *Patti Lu* at Campbell River with herring net and processing hut on deck. The boat was built in Tacoma and is still rigged for table seining with a power block as a typical 58-foot Alaska Limit Seiner. These boats demonstrate why gear limits don't work, as Alaskan fishermen have perfected the use of the powerblock so as to bring fish aboard as quickly as a Canadian drum seiner. The shelter on the side protects the freezer.

In summer 2009, Hutch had fished the aluminum seiner *Cape Knox* for the Canadian Fishing Company after the boat's previous skipper, the legendary Benny Lagos, retired at ninety-two. Canfisco is the same company that Hutch had started out with in Bones Bay over a half-century earlier. But his larger interest was in a herring-bait fishery in partnership with his grandson Marc. He had sold his fine big boats and bought a classic wooden 55-foot Alaskan seiner, the *Patti Lu*. He brought her to Canada and installed freezers. For the past three years, he and a crew from Fort Rupert had been fishing herring in Deepwater Bay on the Quadra Island side just north of Seymour Narrows. Each of the five crew members had a three-ton permit from the Department of Fisheries. They fished a 35-to-40-fathom deep net about 200 fathoms long.

The fish are not concentrated in large schools, but the bait catchers only need a few tons at a time. They put the herring into a live pond and starve them for two or three weeks, then vacuum-pack and freeze them in small lots ready for the sport fishermen. They deliver the fish in Campbell River where a middleman markets them to sport-fishing lodges and stores. "We're not going to get rich but the boys like it," Hutch said. Years ago, some fishermen did a similar fishery and kept the herring in ponds near the top end of Vancouver Island to sell as bait to the outward-bound halibut fishermen. That fishery no longer exists, but as Hutch said to me about commercial fishing in BC, "It's still there if you want to do it."

Hutch has mentored his daughter Christine's thirty-year-old son Marc since he was eleven years old. Now Marc often takes the *Patti Lu* out on his own, though when I visited him on board in January 2010, he said, "It's great to run the boat myself but somehow everything just goes smoother when Grandfather is onboard." Having come

Marc Peeler, Hutch's grandson, is working with the *Patti Lu's* computer. The Wesmar sonar under the computer screen is important in finding herring.

from uncounted generations of fishermen on his maternal side, Marc is a fisherman by both nature and nurture: "I don't think that I've missed a season since I was born." His father, Dale Peeler, became a fisherman after moving out to BC from the prairies. When he was small, Marc spent a couple of seasons out on his father's boat. The rest of the time he has been with his grandparents' boats. "My love of fishing was always there from the time I was a kid making boats out of egg cartons," he recalled.

Marc took time off school in grade ten to fish and then left school to fish full-time in grade eleven. But he has clearly expanded his knowledge base by following both the theory and practice of commercial fishing. One year he even attended a World Fisheries Conference in Portugal with his mother. At nineteen, he worked on the deck of an Alaskan salmon seiner and at twenty he skippered the *Patti Lu* there for the first time. "I was a hot-

A salmon farm north of Fort Rupert, BC.

headed kid and spent too much time chasing the fish." Marc stayed with the Alaskan fishery for five years and in his final year he recalls being stressed going up with a totally green crew. But then he found that he could train them the way he liked, and it proved to be one of his best years.

His time in Alaska also gave Marc an opportunity to learn about another management regime and he was impressed. "When I fished in southeast Alaska," he explained, "I was ashamed to say that I was from BC. It seems that there is a lot more heart in the management up there." Marc continues to go up to Alaska for the Sitka herring fishery each year. "It's more like a holiday because I'm working on deck rather than skippering the boat as I do for the Canadian herring fishery," he said.

In the summer, Marc runs one of the Fort Rupert Band's salmon seiners. To keep busy through the frequent closures, he harvests kelp for a Port Hardy plant that converts it to liquid fertilizer. Together with the winter herring bait fishery and then the spring roe fishery in both BC and Alaska, Marc is finding a means to be a professional fisherman in this new era. With the *Patti Lu* set up for freezing it would make a great prawn boat and he is hoping to find a licence and an experienced mentor to take up that fishery as well. "I really want to get more involved with the Fort Rupert Band's fishery," he said. "I feel privileged to be fishing still and to have learned from my grandfather. Now I want to be more professional." The privilege of learning from a grandfather is great. To have learned from one with the knowledge of Hutch Hunt is huge. Marc and Hutch work side-by-side, sharing a respect for the tides and the sea. With a strong work ethic learned from his grandfather, Marc will carry on the millennium-old Hunt traditions and pass them on to others.

Index